T0189643

Landmarks

Kai-Florian Richter • Stephan Winter

Landmarks

GIScience for Intelligent Services

 Springer

Kai-Florian Richter
Department of Geography
University of Zurich
Zurich, Switzerland

Stephan Winter
Department of Infrastructure Engineering
University of Melbourne
Parkville, VIC, Australia

ISBN 978-3-319-35229-9   ISBN 978-3-319-05732-3 (eBook)
DOI 10.1007/978-3-319-05732-3
Springer Cham Heidelberg New York Dordrecht London

Dedicated to Anja and Elisabeth

Preface

Navigare necesse est
(attributed to Gnaeus Pompeius Magnus)

Landmarks are a fascinating subject of research. On one hand they are ubiquitous in our thinking and speaking about our living environment. We refer to them when describing our internalized image of place or when describing where we are or where we will meet. Tourist brochures try to convey the image of a place by referring to its landmarks Paris is the Eiffel Tower, the Seine river, Notre Dame, and the Louvre. The first-time tourist arriving in Paris may not know more about that place than these landmarks. Experts call this *landmark knowledge* and see landmarks as building blocks of more complex spatial knowledge such as route or survey knowledge. But despite their fundamental role for spatial knowledge, landmarks are surprisingly difficult to define and characterize. As a consequence, they are also difficult to capture in formal models for computation and information services.

To put it bluntly there is a gap between the way how people memorize, think, and talk about their environments and the capacity of information services to store, analyze, or interact with people about environments. A landmark is neither a feature type in current spatial databases, nor do the database taxonomies reveal anything of the nature of landmarks. Let us just check with the above examples of landmarks. Instead of belonging to a single type *landmark* in a spatial database, they actually belong to very different types: in this case tower, river, church building, and museum, or, at higher levels of abstraction, man-made or natural objects. Databases are also restricted by their geometric approach to describe location, while landmarks would require a qualitative rather than quantitative approach. Natural language expressions such as "in front of the Louvre" or "not far from Notre Dame" are hard to interpret by means of a vector geometry used in current spatial databases. A rich number of similar examples can be cited to support that using current services is not intuitive in this sense or, put differently, that these services are not intelligent in the sense of Turing's envisioned intelligent machine [6,7]. And these services reach far into our lives by answering the fundamental *where* and *when* questions in any context. They include services such as simple search, webmapping, mobile location-based services, mobile guides, car navigation services, public transport planners, and emergency call centers, as they involve more and more natural language interaction.

This book will address this gap. It is reporting on latest research on landmarks in geographic environments and practical applications of this research in information service provisions. It covers a spectrum of disciplinary fields encompassed by what has been called cognitive computation by some or spatial cognitive engineering by others. The disciplines involved are from life sciences (neurosciences), social sciences (psychology, cognitive science, linguistics), engineering (artificial intelligence, information systems), physical and mathematical sciences (geomatics), and humanities (geography, philosophy).

The reader can expect from this book a broad **scope** covering perceptual and cognitive aspects of natural and artificial cognitive systems, conceptual aspects of trying to define and formally model these insights, computational aspects with respect to identifying or selecting landmarks for various purposes, and communication aspects of human-computer interaction for spatial information provision. The **origin** of the book goes back to our own, originally separate work on computational issues of landmarks, which started about a decade ago, perhaps kicked off by one particular, frequently cited paper [4]. Of course, landmarks were by then already a well-studied topic in spatial cognition research (e.g., [2, 3, 5]). A rich body of work had been developed, which we will present in this book in a systematic manner, including outlining the still open questions. Portions of the presented ideas led also to the world-first commercial navigation service using landmarks selected based on cognitive principles [1].

Accordingly, the **purpose** of this book is to provide a review of this line of research, structured into cognitive, conceptual, computational, and communication aspects. This is in particular valuable because it represents a synopsis of research in different disciplines and thus not only addresses a breadth of topics but also bridges between different traditions of thinking. It is also timely since the research in these four areas has reached levels that allow for a first time a consistent synopsis.

The **intended audience** of this publication are certainly graduate or postgraduate students—they will profit from a compact reader summarizing and synthesizing a large number of research papers—but beyond that also the interested public, from the enthusiastic geeks maintaining crowd-sourced datasets like OpenStreetMap, over the early adaptors of novel tools such as navigation services, to the curious people being interested why certain things are so hard for computers ... such as thinking about our environments like we humans do.

Zurich, Switzerland   Kai-Florian Richter
Parkville, VIC, Australia   Stephan Winter
March 22, 2014

References

1. Duckham, M., Winter, S., Robinson, M.: Including landmarks in routing instructions. J. Location-Based Serv. **4**(1), 28–52 (2010)
2. Lynch, K.: The Image of the City. The MIT Press, Cambridge (1960)
3. Presson, C.C., Montello, D.R.: Points of reference in spatial cognition: stalking the elusive landmark. Br. J. Dev. Psychol. **6**, 378–381 (1988)
4. Raubal, M., Winter, S.: Enriching wayfinding instructions with local landmarks. In: Egenhofer, M.J., Mark, D.M. (eds.) Geographic Information Science. Lecture Notes in Computer Science, vol. 2478, pp. 243–259. Springer, Berlin (2002)
5. Sorrows, M.E., Hirtle, S.C.: The nature of landmarks for real and electronic spaces. In: Freksa, C., Mark, D.M. (eds.) Spatial Information Theory. Lecture Notes in Computer Science, vol. 1661, pp. 37–50. Springer, Berlin (1999)
6. Turing, A.M.: Computing machinery and intelligence. Mind **59**(236), 433–460 (1950)
7. Winter, S., Wu, Y.: Intelligent spatial communication. In: Navratil, G. (ed.) Research Trends in Geographic Information Science, pp. 235–250. Springer, Berlin (2009)

References

Acknowledgements

This book presents the work of many people to whom we are indebted for their inspirations, their collaborations at some stage of our pathways, or for challenging our ideas in discussions or in anonymous reviews.

Our own work was also supported by a number of research funding agencies, including the Australian Research Council (*Talking about Place*, LP100200199), the Go8/DAAD exchange program (*Cognitive Engineering for Navigation Assistance*), the Institute for a Broadband-Enabled Society (*Crowd-Sourcing Human Knowledge on Spatial Semantics of Placenames*), the German Research Foundation DFG (*Transregional Collaborative Research Center SFB/TR 8 Spatial Cognition*), the Australian Academy of Science (*Smart Interaction for Wayfinding Support*), and the Swiss National Foundation (*Computational Methods for Extracting Landmark Semantics*).

Acknowledgements

Contents

Chapter 1
Introduction: What Landmarks Are, and Why They Are Important

Abstract Landmarks, by their pure existence, structure environments. They form cognitive anchors, markers, or reference points for orientation, wayfinding and communication. They appear in our sketches, in descriptions of meeting points or routes, and as the remarkable objects of an environment in tourist brochures. With all their significance for spatial cognition and communication, landmarks pose a major challenge for artificial intelligence and human-computer interaction. So far, research aiming for intelligent interaction design has suffered from a lack of understanding and formal modelling of landmarks. Here we will clarify our words' meaning and draw some boundaries of our discussion, in preparation for integrating landmarks in artificial intelligence applications.

1.1 What Landmarks Are

What hasn't been called a landmark so far! We speak of a *landmark development*, a *landmark victory*, a *landmark court decision*, a *landmark resort*, a *natural landmark*, call a skyscraper a *landmark*, or the Eiffel Tower (Fig. 1.1), or the Sydney Opera House. The term is not only used as an attribute, but also as a name such as the *Landmark Tower*, Japan's tallest building,[1] or *Landmark*, a rural settlement in Manitoba, Canada.[2] Wikipedia[3] defines a landmark "including anything that is easily recognizable, such as a monument, building, or other structure." Particularly useful for the purpose of this book is the reference to *recognition* in this definition. Cognition and embodied experience will play a significant role in our exploration

---

[1]http://www.yokohama-landmark.jp/web/english/, last visited 23/12/2013.

[2]http://www.geonames.org/maps/google_49.672_-96.822.html, last visited 23/12/2013.

[3]http://en.wikipedia.org/wiki/Landmark, last visited 23/12/2013.

K.-F. Richter and S. Winter, *Landmarks: GIScience for Intelligent Services*,   1
DOI 10.1007/978-3-319-05732-3_1, © Springer International Publishing Switzerland 2014

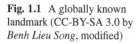

Fig. 1.1 A globally known landmark (CC-BY-SA 3.0 by *Benh Lieu Song*, modified)

of landmarks. But then this definition is also somehow imbalanced, first speaking of 'anything' in the intensional part, and then mentioning only examples taken from the built environment in the ostensive part.

Let us have a look into another dictionary. Merriam Webster distinguishes three meanings: "(1) an object or structure on land that is easy to see and recognize, (2) a building or place that was important in history, or (3) a very important event or achievement[4]."

- Meaning 1 is more or less covering the previous definition, although it is questionable why landmarks should be restricted to land. Landmarks arguably exist, for example, also on the ice cap of the North Pole, indoors, and even at sea: Navigators have used Polaris as a landmark for centuries. But importantly Meaning 1 refers again to cognitive processes based on embodied experience, which is what we will build upon later. Let us ignore that the wording requires 'seeing' in addition to 'recognizing'. If seeing would be required blind people would have no concept of landmarks, or soundscapes would have no landmarks, which is both not true. However, the embodied experience postulated above is expressed here in the reference to locatable objects or structures.

---

[4]http://www.merriam-webster.com/dictionary/landmark, last visited 23/12/2013.

Fig. 1.2 A landmark: an object made to mark out a location or boundary on land

- Meaning 2 adds time as an essential factor to embodied experience: Landmarks are not only what 'is easy to see and recognize' but also what has been 'easy to see and recognize' in the past and therefore still can be located by memory.
- Meaning 3, relating to events and achievements, is of a different nature. Despite the fact that events and achievements can be located in principle—everything happens somewhere, even a Black Friday or Global Financial Crisis was produced somewhere—the relationship to a location is now taking a back seat, and what is now 'easy to see and recognize' is answering a *what* question rather than a *where* question. With this shift of focus comes along a mapping of the concept from its source domain of physical objects located in space to a target domain of temporal objects [15]. This means Meaning 3 is taking the concept of landmark rather metaphorically [26], preserving the characteristic of being cognitively salient, and thus, structuring memories.

Etymologically, the English word landmark, or *land* and *mark*, comes from an old form *landmearc* ("boundary") or *landgemirce* ("boundary, limit, frontier"), and can also be found in German, Danish and Swedish. (Polygonal) boundaries are constructed by setting out the corners (Fig. 1.2). Hence, the cornerstones bear all the meaning of the boundary, or are re-*mark*-able.

Sticking for a while with "an object or structure on land that is easy to [...] recognize", WordNet [9, 32], which is not a dictionary in the strict sense of the word, but rather a semi-formal ontology or lexicon of concepts, provides us with the following conceptual hierarchy of the term:

Level 0   *Landmark*—(the position of a prominent or well-known object in a particular landscape; "the church steeple provided a convenient landmark")

Level 1   *Position, place*—(the particular portion of space occupied by something;
"he put the lamp back in its place")

Level 2   *Point*—(the precise location of something; a spatially limited location;
"she walked to a point where she could survey the whole street")

Level 3   *Location*—(a point or extent in space)

Level 4   *Object, physical object*—(a tangible and visible entity; an entity that can
cast a shadow; "it was full of rackets, balls and other objects")

Level 5   *Physical entity*—(an entity that has physical existence)

Level 6   *Entity*—(that which is perceived or known or inferred to have its own
distinct existence (living or nonliving))

This hierarchy of hypernyms of *landmark* tells us something about the conceptual
space of the term *landmark*. Each level inherits from the more abstract concepts and
adds its own specific meanings. Especially the derivation from place and location
suggests a formal property of an image schema of contact to a place (e.g., [11, 25]),
a meaning that is reflected in below's approach to define landmarks as references to
locations.

1.1.1 An Extensional Approach

Theoretically, one approach to capturing the meaning of a concept is by forming
a *taxonomy*, and enlisting all entities that belong to a particular category in this
taxonomy. So let us assume our taxonomy—whether a cognitive or a database
taxonomy does not matter—contains a category *landmarks*. Is it then humanly
possible to enlist all landmarks, or, with other words, can the meaning of landmarks
be captured by providing a comprehensive list of all landmarks?

We could think of landmarks in our hometown and start a list. Asking friends
would add to the list. But asking friends of friends would add further, and may
bring up objects we were even not aware of. Are these all landmarks? Perhaps we
would also like to challenge some of the elements picked by others. So in order to
answer the question whether an extensional approach is the way forward the proper
response is: The approach is feasible, but unsatisfying. Our favorite search engine
may have such a list and use it on their webmaps, but there is no explanation in such
a list why these elements were considered to belong to the category, and others
are left out. Correspondingly, there may remain disagreement about individual
elements. The deeper issue with an extensional approach is about completeness. In a
closed world assumption only listed objects are landmarks. But there is no known
procedure guaranteeing completeness as long as the selection criteria are not made
transparent. In our dynamic world a related issue is maintaining the list by adding
and removing items.

Having said this, current approaches of user-generated database content do apply an extensional approach. Letting users add what they think is a landmark may not explain what a landmark is, but it has a potential to collect datasets of a global scale, as has been proven with OpenStreetMap.[5]

1.1.2 A Cognitive Semantics Approach

Rosch [38] argues that two principles drive the formation of categories in the mind. One is cognitive economy, calling for grouping similar things together and giving them a name. The other principle is rather a recognition that the continuous world outside of the body is actually structured and forms natural discontinuities. Categories are then, economically, formed by objects that have many attributes in common and share not many attributes with members of other categories [39].

This idea has to be taken by a grain of salt, though. Wittgenstein had famously interjected that at least several natural categories appear to have no common attributes, or none that are not shared with members of other categories—he used the category of *games* (board games, card games, olympic games) for this argument [50]. Instead of commonalities Wittgenstein argues for a *family resemblance*, which is only a matter of similarity. As a consequence, Rosch [38] suggests to apply a measure of category resemblance derived by Tversky from similarities: "Category resemblance is a linear combination of the measures of the common and the distinctive feature of all pairs of objects in that category" ([47], p. 348). Thus, category resemblance describes the inner coherence of a category, without requiring a catalog of shared attributes. Imagine a space spanned by some conceptual properties. In this space each object forms a node. Convex categories can be formed by clustering these nodes such that if x and y are member of a category and z is between x and y than z is member of the category as well [16]. These conceptual spaces form a mathematical basis to express family resemblance.

Extending such a notion of a category, prototype theory calls in more central entities—*prototypes*—and accepts graded membership to a category [27, 37, 38]. For example, probably most people would agree that the Eiffel Tower is a landmark, and in surveys the Eiffel Tower is always highly ranked (e.g., Fig. 1.3), which makes the Eiffel Tower a prototype in the category of landmarks. But how many people would agree that the blue house at the street corner is a landmark, the ATM in the mall, the T-intersection where one has to turn right, or the only tree far and wide marking the entrance to the farm? These are perhaps more ambiguous entities in the category.

By the way, landmarks are not special in this respect. Any classification of objects in geographic space is to some extent arbitrary, and has its prototypes and its boundary cases. For example, we may have a clear understanding of what a building is, a mountain, or a road. They are everyday terms, and basic categories in

---

[5]http://www.openstreetmap.org, last visited 3/1/2014.

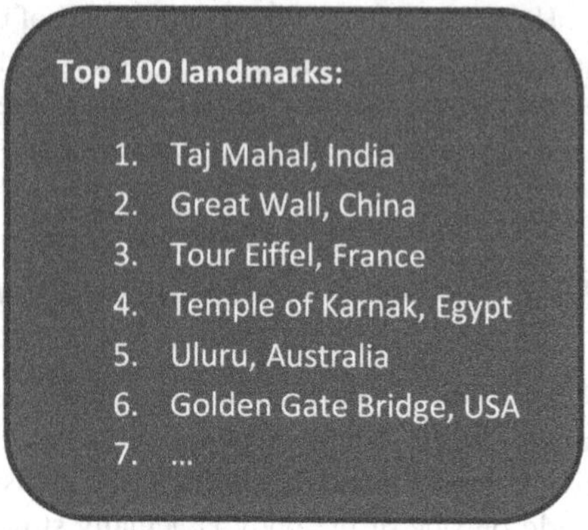

Fig. 1.3 A ranked list of
landmarks

Top 100 landmarks:

1. Taj Mahal, India
2. Great Wall, China
3. Tour Eiffel, France
4. Temple of Karnak, Egypt
5. Uluru, Australia
6. Golden Gate Bridge, USA
7. ...

the context of geographic information [39]. They are standard elements in spatial
databases: A typical GIS contains database layers of (representations of) buildings
and roads, and gazetteers of geographic names contain representations of mountains.
According to prototype theory each of these categories will have typical examples
as well as less clearly assigned entities. When it comes to these boundary cases in
categorizations, people start to disagree or become uncertain. Is a single garage, a
shed, or a kennel still a *building*? Is a laneway, a mall, or a trek still a *road*? Is a hill
a *mountain*? Is Uluru a hill or a mountain? These questions are important for map
makers and database administrators alike since they decide about the cleanliness of
their products. A producer does not want to have a product with information many
people disagree about. These questions also decide about the ability to compare
different map sources or databases. Map updating, for example, should not merge
datasets of varying semantics. It has even been shown that these classifications can
vary across languages and cultures [29, 30]. Thus, there is no definite answer to the
question which object is a landmark and which is not. Landmarks are countable but
are not finite.

There are other reasons adding evidence to this conclusion. For example, the
world is constantly changing, and over time it can change whether a geographic
object is a landmark or not. The first skyscraper in Chicago was a landmark, at
least for some time, until it became one of many and others were more outstanding.
Accordingly, classifications of objects can be made only for a certain time. But even
at a snapshot in time we have already seen it is impossible to provide a complete
list. Now we know why. It is because of graded membership. Additionally this grade
of membership to the landmark category depends also on the context. For example,
a city looks very different by night than by day, and landmarks in night scenes
may be quite unimpressive in daylight. Or consider a café that became special for a
couple because it is the place where they first met. It is one of the locations in the
city they refer to when they explain other locations to each other. For the two of them
it has become a landmark they share even if this is one of many cafés in that street.
For others, especially those who never visited the café, it is not. The New York City
Apple Store (Fig. 1.4a) may have the meaning of a landmark for people of some

Fig. 1.4 Landmarks: (**a**) New York City Apple store; (**b**) Sydney opera house

interests and particular age, but not for others. The Sydney Opera House (Fig. 1.4b) may be a prime example of a globally recognized landmark, but its cleaners may see this labyrinth of a complex building differently, on another spatial granularity, and perhaps with other landmarks within for their own orientation and communication purposes ('the box office', 'the stage', 'Bistro Mozart'). Thus, being a landmark is not a global characteristic of an object, but a function of parameters such as the individual that perceives and memorizes an environment, the communication situation, the decision at hand, and the time. The latter argument means that even prototypes cannot be considered prototypes in all cases, since there might be no such thing that is always, i.e., in any context a well-suited landmark.

1.1.3 An Intensional Approach

With only limited successes of extensional and cognitive semantics approaches, perhaps an intensional approach to capture the concept of landmarks is more promising, specifying the necessary and sufficient conditions of belonging to the category *landmark*. In order to relate to people's (in principle, any cognizing beings') embodied experience and cognitive processing of their living environments the following intensional definition will be chosen for this book:

> **Definition:** Landmarks are geographic objects that structure human mental representations of space.

Other intensional definitions appear to be dependent on this one. For example, Wikipedia's "anything that is easily recognizable" (Sect. 1.1) has this cognitive flavor built in if recognition is a match of a perception with a mental representation. The more generic "Landmarks are prominent objects in the environment" can be understood in two highly correlated directions. One refers to perceivable properties in the environment; accordingly a landmark is standing out by some properties, or generates an experience for an individual that contributes to the mental representation of this environment and can be recalled. The other direction refers to the degree an object is part of common knowledge. Since every person depends on their human senses the likelihood is high that an object's outstanding properties are perceived by many. Thus, landmarks are suited for communication of environmental knowledge. Sorrow and Hirtle's much cited definition [43] is of this sort, reading effectively: "Landmarks are geographic objects of outstanding visual, cultural or structural properties". Their experience must structure human mental representations of space.

Even the extensional approach behind Fig. 1.3 follows the intention of this definition. In order to illustrate the effect, consider the following experiment: Take the six landmarks ranked in the figure and order them geographically on a blank sheet of paper. You will find arranging them not too difficult. What you have actually achieved is an externalization of your mental spatial representations. The same experiment can be easily translated into the context of your own hometown. Imagine a number of prominent objects in your hometown, and ask your friends to arrange these objects for you spatially. Again, they will be able to produce a sketch reflecting satisfactorily the outlay of your hometown.

In a very similar experiment by Lynch [28] people were asked to sketch their hometown. Comparing a large number of sketches, Lynch found commonalities in the structure of the sketches. One common element in these sketches he called *landmarks*,[6] or, in his understanding, identifiable objects which serve as external reference points.

Our definition of landmarks is purely functional: being a landmark is a role that objects from any category can play. It emphasizes that landmarks are mental constructs. In alignment with Meaning 1 and 2 from above it covers for objects that stand out in an environment such that they have made (or can make) an impression on a person's mind. This experience is not limited to visuo-spatial properties of the object itself, nor to current properties or to shared experiences. Counterexamples would be "the café where we met", "the park where we had this unforgettable picnic" or "the intersection where I was nearly killed by a car". It is also not limited to human-made things, as the definition by Wikipedia above might suggest; it can equally well be a natural object such as the Matterhorn (Fig. 1.5), a widely visible mountain in the Swiss Alps with a characteristic shape, and hence frequently used for localization and orientation. But in their function landmarks must have certain properties, most importantly: being recognizable in the environment.

---

[6]We will later argue that all elements Lynch has identified can be considered as landmarks.

Fig. 1.5 The Matterhorn, a
landmark in the Swiss Alps,
on a historic photograph

However, the chosen definition is not unproblematic. For a scientist the definition remains questionable just because what happens in human minds is not directly observable or accessible. So if we agree that there are mental representations of what is in the world outside of our bodies, then indirect methods must be accepted to reconstruct the elements and functioning of these representations. These indirect methods are the toolboxes of cognitive science, with a strand in the neurosciences, studying the structure and work of the senses and the brain, and another strand in cognitive psychology, studying responses or behaviours of people, such as learning, memorizing, language, or spatial behavior. This cognitive perspective on landmarks could easily fill a book by itself. Therefore, we will keep the discussion of the cognitive aspects concise, and instead focus on how the understanding of the role and working of landmarks in spatial cognition and communication can inform the design of smarter systems interacting with people.

It must also be said that this definition is in conflict with other common understandings of landmarks. For example, Couclelis et al. have called *anchor points* what we have called landmarks [4]:

"Much of the work in spatial cognition has focused on the concept of imageability. In brief, this assumes that there will be elements in any given environment (natural or built) which by virtue of their distinctive objects (for example, form, color, size, visual uniqueness), or by virtue of some symbolic meaning attached to them (places of historic importance, of religious or socio-cultural significance, etc.), stand out from among the other things in the environment. Because such elements are outstanding, literally, they are likely to be perceived, remembered, and used as reference points by a large number of people in that environment. This is the notion of *landmark* as popularized by Lynch's seminal work on the 'Image of the City'. Anchor-points (anchors for short) are closely related to landmarks, both concepts being defined as cognitively salient cues in the environment. However, as represented in the literature, landmarks tend to be collectively as well as individually experienced as such, whereas anchors refer to individual cognitive maps. Although one would expect to find several local landmarks among the anchors in a person's cognitive map, many anchors (such as the location of home and work) would be too personal to have any significance for other, unrelated individuals. Further, landmarks are primarily treated as part of a person's factual knowledge of space, whereas anchor-points are supposed to perform in addition active cognitive functions, such as helping organize spatial knowledge, facilitating navigational tasks, helping estimate distances and directions, etc. Finally, landmarks are concrete, visual cues, whereas anchor-points may be more abstract elements that need not even be point-like (e.g., a river or a whole city in a cognitive map at the regional level)" (p. 102).

In fact, our definition does not make such a distinction between anchor points and landmarks. Instead, we argue that, in principle:

- any property standing out in an environment can be shared with another person, and
- that any sharing of experiences is limited to (larger or smaller) groups.

For example, 'my home' is shared with a few people, and 'Eiffel Tower' is shared with many people. But even the Eiffel Tower is not known by the universe of all living people. Furthermore, experience can be lived and communicated. Landmarks can be learned from text, from maps, or from conversation (e.g., somebody may tell me: "At that intersection I was robbed", which attaches an outstanding memory to this object). Thus, all what is required in a communication situation is taking a perspective. As people do adapt themselves to their communication partner in their choice of landmarks, so must the machine. This involves a capacity for context-aware computing [6]. Different groups (e.g., family, colleagues, people living in the eastern suburbs, tourists) share different sets of anchor points, or can expect that certain landmarks can be experienced by particular individuals.

Relating the notion of landmarks to embodied experience has some tradition in research in spatial cognition. In environmental psychology, for example, Siegel and White wrote, "landmarks are unique configurations of perceptual events (patterns). They identify a specific geographic location" ([42], p. 23), and similar words can be found elsewhere (e.g., [8]). According to their distinctive experience, landmarks should be "the most easily recalled attributes of a region" [41]. Sadalla et al. then go on to "explore the function of landmarks as spatial reference points, points that serve as the basis for the spatial location of other (nonreference) points" (*ibid.*). Similarly the definition has been supported by neuroscience, which has shown that objects relevant for navigation and orientation do not only engage object recognition in the brain but also areas associated with spatial memory (e.g., [21, 24]).

Presson and Montello use this definition as well ("objects that are relatively better known and define the location of other points") in their discussion of the nature of landmarks [35]. In particular they point out that landmarks, in order to be able to define the location of other points (objects in our terminology), must be distinct from these other elements in spatial memory, and central to the nature and organization of mental spatial representations. We will come back later to a discussion that landmarks may be stronger or weaker in their distinct experience, and that stronger ones may be used as reference points to locate weaker ones. In this regard, only objects that are located by reference to "better known" objects are not landmarks. Presson and Montello continue, referring to the observation of asymmetric distance estimates between reference and non-reference points made by Sadalla et al.: "The relation of reference to non-reference points is assumed to be asymmetric although the notion that there are a few elements to which many others are spatially related does not require this to be so. Non-reference points are more likely to be defined in terms of their relation to reference points than vice versa, and the judged distance from a reference point to a non-reference point may not be equal to the same distance judged in the reverse direction". Landmarks, to be used as points of reference, must be objects that are relatively better known than the other objects in their neighborhood. In this regard, Appleyard [1], the urban designer, had already postulated: "We have to go beyond Lynch's identification of known urban element types. We must determine the reasons why these elements are known which means discovering the attributes that capture attention and hold a place in the inhabitant's mental representation of his city" (p. 131). But generally in urban design and architecture the notion of landmarks is restricted to the build environment, or more precisely buildings, and falls too short for the purpose of catching human mental spatial representations or, correspondingly, intelligent spatial systems.

1.1.4 An Artificial Intelligence Approach

Prototypes allow an ostensive capture of the meaning of a category, i.e., by example. Someone may explain the concept of landmarks this way: "Landmarks are, for example, the Eiffel Tower, the Great Pyramid of Giza, Taj Mahal, Uluru, the Great Wall, and Golden Gate Bridge." Ostensive approaches are not uncommon. Think of the "top 100" lists on the internet (Fig. 1.3) not in an extensional but ostensive way. Similar lists of examples can be collections of personal favorites, lists of travel promoters ("must-do's"), crowdsourced lists (e.g., surveys), or lists produced from data mining (e.g., a search engine's number of hits, or your favorite travel website's lists ranked by reviews submitted by travellers). These lists tend to produce the more agreeable, less ambiguous entities of a class.

On 28 June 2013, Kate Schneider, travel editor of News Ltd., wrote: "Kings Park War Memorial in Perth. Fremantle prison. Melbourne's Block Arcade. What do these places have in common? Well, they've all made a list of the nation's best landmarks by travel website TripAdvisor. Yep, really. According to the site, they are among the top spots to go in Australia for "enriching and entertaining experiences". The list was based on millions of reviews submitted by travellers on the site over the past year. It also includes iconic Aussie attractions such as the Sydney Opera House and the Sydney Harbour Bridge, mixed in with grim locations such as the Port Arthur Historic Site. The list had us wondering, is this the best [Australia has] to offer the world?"

http://www.news.com.au/travel/australia/australia8217s-top-10-landmarks-named-in-tripadvisor-list/story-e6frfq89-1226670801261#ixzz2XjmExZCp, last visited 3/1/2014.

Machine learning algorithms [33], one of the core tools of artificial intelligence [40], can use such lists as training data and then try to identify more landmarks. The task of machine learning (or any learning, in that respect) is not trivial: If the examples in Fig. 1.3 above are landmarks, does then the Alcazar in Spain count as a landmark as well? In order to evaluate such a question, an algorithm (as well as a human mind) would collect characteristics of the given landmarks and search for patterns of similarity for conceptual resemblance.

Lists, or datasets of candidates, however, do not resolve the challenges discussed before. These lists require meta-level descriptions of the context of their validity, a link between ranking and grading, local structures by spatial relationships, and they require continuing maintenance due to changes in the world. Most critically, however, the relationships between the elements of the dataset and the definition of landmarks must be established methodologically. So just as cognitive science is challenged identifying the nature and role of landmarks, artificial intelligence is challenged as well. Generally, spatial databases are fundamentally different, and incompatible with mental spatial representations. Analysis in spatial databases is based on geometric representations and geometric algorithms. For example, routing in navigation systems will apply an exact shortest path algorithm, such as Dijkstra's [7]. People's mental spatial representations do not allow this kind of reasoning. Their mental representations are good at (fast) guesses, but not at exact computations [18, 22].

Consequently, their language is quite different from autonomous systems such as robots. While a robot is happy with instructions such as "Move in direction of 354.6 degrees for 35.7 meters, then turn by 32.6 degrees", a human would find this hard to realize, let alone to manage the cognitive load with memorizing all these numbers (imagine a full route description of this sort). People, in contrast, communicate in ways the machine has difficulties to interpret. For a machine, understanding "To the café at the library" requires a number of tasks, such as identifying *the library* in a spatial dataset by interpreting the conversational context, and then disambiguating the *café at the library* from all other cafés. The latter involves searching current business directories, an ontological matching of all businesses that can count as a *café*, and last but not least an interpretation of the preposition *at*, a qualitative spatial relationship that is vague and underspecified. Judging from the structure of this phrase, where the café is located relative to the library, the library appears to be a landmark in the sense of the definition above. Thus, the capacity to interpret landmarks is essential for smart human-computer interaction.

And once the machine succeeds with the interpretation of the phrase, and computes a route for this person, this person expects human-like directions, for example, "Turn right at the traffic lights". The construction of these directions in natural language is as difficult for a machine as the prior challenge of interpreting natural language. It requires again a number of tasks, such as applying a smart concept of *right-turn* in urban streetscapes, which not always show rectangular intersections, and referring to suited landmarks such as, in this case, the *traffic lights*, but not the trash cans, fire hydrants, or the faceless office buildings located at the intersection. Thus, the capacity to generate landmark-infused spatial language is also essential for smart human-computer interaction.

This, in a first instance, requires understanding of what a landmark is, i.e., formal models, data, and algorithms for reasoning.

1.2 Related Concepts

From the definition and discussion above it becomes clearer that landmarks emerge in the process of perceiving, learning and memorizing an environment in a particular context, and that these memorized landmarks will be picked up in spatial reasoning or communication processes. For objects in the environment to acquire landmark quality these objects must somehow stand out. Furthermore, to contribute to the embodied experience of the environment in which a person moves these objects must be related to the human body and human senses. Using a classification relative to embodied experience introduced by Montello [34], these objects have to be identifiable objects in *vista space, environmental space*, or *geographic space*. Vista space is the space covering all objects that can be seen from a single viewpoint, and with the naked eye. Examples are a room, an open plaza, or any other vistas in streetscape. Environmental space is the space learned by locomotion—the movement of the body coordinated to the proximal surrounds—and the integration of this embodied experience. Examples are buildings that are learned only by walking through, or city districts that are learned by walking or driving around. All body senses, including sight (vistas), contribute to an integrated, coherent mental representation of these spaces. Geographic space is the space larger than environmental space such that it can be learned only from symbolic representations such as maps. Examples are countries or even larger cities, which cannot be explored completely by foot, car or another form of locomotion.

With these integrated layers of human experience and mental representations, landmarks will be found at each level. Orientation is helped by outstanding objects in vista space ("the keys are on the *table*"), in environmental space ("the *library* is around the corner"), as well as in geographic space ("*Cologne* must be in this direction"). Having this more graphic image of landmarks at hand, it becomes easier to draw boundaries around the concept of landmarks, and to discuss how other things relate to landmarks. This way, objects that stand out on other than vista, environmental or geographic scale are not considered to form landmarks.

For example, markers on a genome may be important elements to conceptualize and structure their spatial representation. But even if science speaks of 'mapping the genome', and the scientists structure their mental representations of the genome by these markers, mental representations of space at this scale are not directly accessible to the experience of human senses.

It is likely that animals also experience landmarks as geographic objects structuring their mental representations of space. Rich neuroscientific and cognitive literature suggests that *spatial* abilities are among those where insights learned from animals can be transferred to human cognition. However, the animals' embodied experience of environments is different. For example, ants, honey-bees or rats—thoroughly studied animals for their orientation and wayfinding skills—navigate at completely different scales. Hence, this book will generally not expand on animal spatial cognition.

Robots are another domain of knowledge where landmarks play an imminent role. Robots are capable of simultaneous localization and mapping (SLAM, e.g., [44]), learning an unknown or updating a known environment through their sensors. For them landmarks are stable and easily identifiable external points of reference, sometimes artificial tags put out in the environment, sometimes objects of the nature we consider in this book. Since machine sensors, machine mapping and machine reasoning are qualitatively different from human embodied experience, memorization and reasoning, this book will also not address landmarks in artificial systems.

However, in addition to the use of landmarks in other domains we are also going to differentiate landmarks from related concepts, especially where terms are sometimes confused in common language. In the following we discuss *places*, *points of interest*, *icons*, and *metaphorical landmarks*.

1.2.1 Places

Landmarks are points of reference in mental spatial representations. Their function in mental representations is to locate other objects. This function establishes a connection to *place*, which is another geographic concept structuring space, and perhaps one that is even more elusive than landmarks [19, 35].

This elusiveness of place emerges from the tension between the informal world of human cognition and discourse that deals with objects on one side, and on the other side a continuous physical world "out there" that does not come by objects in the first instance [10]. In addition the term has been used with different meanings. Gärling et al. ([17], p. 148f) said: "Place is a concept that is rich in meaning and difficult to define", and that is only rephrasing Aristotle: "The question, what is place? presents many difficulties" ([2], IV/1).

However, in a book on landmarks we can be more specific. A purely philosophic position was taken by Aristotle, who says later in his *Physics* ([2], IV/4):

"What then after all is place? The answer to this question may be elucidated as follows
[...]:

- Place is what contains that of which it is the place.
- Place is not part of the thing.
- The immediate place of a thing is neither less nor greater than the thing.
- Place can be left behind by the thing and is separable.
- In addition: All place admits of the distinction of up and down, and each of the bodies is
  naturally carried to its appropriate place and rests there, and this makes the place either
  up or down."

The last item seems to indicate that places have gravity, an understanding shared
by contemporary geography [3]. In the Aristotelian notion of place objects have
their unique *immediate* place, or footprint: "Hence we conclude that the innermost
motionless boundary of what contains is place" ([2], IV). Two phenomena cannot
have the same immediate place, and one phenomenon cannot be at two different
immediate places at the same time. Beyond the immediate place a phenomenon is
simultaneously nested in an unlimited number of other places. For example, while
I am sitting *at my desk*, I am at the same time *in my office*, *in the department*, *et
cetera*.

However such a philosophical perspective does not yet consider the more
experiential perspective of geography, with its notion of a sense of place (e.g., [31,
36, 45]). A geographer's approach might be calling any meaningful spatial configuration of shared affordances to the human body a place. Such a definition shows
some similarities to landmarks, for example, both requiring perceptual wholeness
and cognitive salience, and both being context-dependent. But it also explains the
difference: Landmarks act as anchor points, and hence are conceptually abstracted
to nodes where no internal structure is required. Their purpose is fast reasoning or
efficient communication. Place, in contrast, captures the meaning and affordance of
a scene, and hence, is rich in structure and complex to communicate. With regards
to the latter, and especially focusing on the context dependency, Freksa and Winter
have pointed to the principle of sufficiency ([49], p. 32):

"Cognition of and communication about place in spatial environments is a matter of
sufficiency. Sufficiency can be captured by contrast sets, that is, by specifying the meaning
of a place in a given context by explicating the contrast to other places. People conceptualize
a portion of an environment as a place if their embodied experience of this portion
shows a wholeness against the background, i.e., if it has some contrasting properties to
its environment or to other places. Referring to such human embodied experience, our
arguments and examples will focus on geographic places, i.e., on places in our physical
environment of vista, environmental, or geographic scale as defined by Montello [34]."

In the same context:

"In cognitive and communicative tasks place functions primarily as spatial anchor. The
anchor locates, and connects by location objects and events as they occur in what-, why-,
when-, or how-constructions. For example, in an expression *An accident happened on
Beaver Street this morning*, the place *on Beaver Street* is anchoring an event in space,
answering a hypothetical or real where question. Places are typically determined by entities
in the geographic environment (objects or events) or by relations between entities in the
environment rather than by externally imposed coordinates and geometric properties."

| Tate Modern Gallery | London, UK | +44 2078878888 | -0.09931, 51.50701 |

Fig. 1.6 POI database entry for the Tate Modern

"As interrelated processes, cognition and language make use of places. For example, spatial reasoning happens on qualitative spatial relationships between places [12], and everyday language refers to named and unnamed places (e.g., *on Federation Square, at the road intersection*) and the relationships between them (e.g., *at Birrarung Marr near Federation Square*). Sketches, as non-metric graphical externalizations of cognitive or verbal representations, also reflect configuration knowledge of places and their relationships. And yet, despite recent progress in neuroscience and cognitive science our knowledge about cognitive representations and reasoning is not sufficient to formally characterize the entities, relations, and operations that would enable us to build a system that reflects the computational processes of spatial representations in our mind."

Whichever perspective is taken, the philosophical or the geographic, place covers a different meaning to landmarks. From the philosophical perspective, every object has a place, but not every object is a landmark. From the geographic perspective, places can also function as landmarks, but are then stripped of their rich internal meaning and instead linked with a location. More specifically, since places are inhabited, their inside or internal structure must be important. Places are related to being, or, as Tuan called it, *resting* [45]. Since landmarks are points of reference to locate other objects, their outside must be important—as they are perceived or imagined by a person linking two objects. Landmarks are related to movement, such as passing by (waypoints), turning (decision points), or heading (distant points of orientation).

1.2.2   Points of Interest

Some geographic information systems contain *points of interest* databases, and may use this terminology, which is sometimes abbreviated to *POI*, also in their user interface. For example, car navigation services and public transport trip planners offer users to specify their destination by selecting a point of interest from their database, and mobile location-based services provide points of interests on their you-are-here maps. A point of interest is simply a point (typically a point on a map, or a GPS coordinate in some spatial reference system) that somebody by some authority has declared to be interesting. A point of interest typically comes with a name, carrying the semantics of what can be found at the location characterized by this point (Fig. 1.6).

Yet what can be considered to be interesting depends on the particular context a person may find themselves in. As a car driver, they will find gas stations, parking houses, and speed cameras interesting. As a tourist they will find museums, churches, restaurants and also hotels interesting. As a public transport user they will find stations, event locations, or public institutions interesting. We got used to specialized services catering for these markets, such as car navigation systems or tourist guides. However, what about generic information services? Once we are able to take our car navigation system out of the car and use it for pedestrian navigation these systems would need mechanisms to determine the context of a particular user, or query, to come up with a relevant result. But then, an economic geography researcher will find perhaps living places and workplaces interesting, which are typically lacking in point of interest databases. Concerned parents want to know where their children are, and points representing these locations are also lacking in point of interest databases. Or a serendipitous armchair traveler might enjoy to discover locations of objects of unexpected categories, which for a system are even harder to predict than profiled preferences. Thus, prefabricated and stored points of interest come along with some paternalism: Something at some location has been deemed to be interesting by somebody in a specific context.

Are points of interest landmarks? As indicated, current systems choose particular categories of points of interest for particular contexts, for example, car driving or using public transport. They do consider a potential service for the user, such as navigating by a given means. This can include communicating destinations, presenting waypoints that may service the current mode of travelling, or presenting choices of routes for a given mode. Since points of interest are service-oriented, they are also attractive advertising tools. For example, particular brands may want to ensure that they are present on car drivers' maps. From their perspective it might be even more attractive to block other brands from being listed. But despite the commercial flavor and potentially compromised selection of points of interest there is another substantial argument why points of interests are not landmarks: The selection of points of interest does not consider the appearance to the human senses, and more generally, does not aim to support human orientation and wayfinding. Some of the point of interest categories cannot be expected to be easily recognized from outside, or to stand out in their neighborhood. For example, the POI category of medical doctors may produce POIs on a map, but they will actually be hard to find in the environment for a car driver, with practices being unremarkable from the outside (according to professional codes of the profession). They also can be located in hidden places such as in malls, or above ground levels of buildings. But if the doctor's practice does not stand out in its neighborhood, it will not structure mental representations because people cannot experience it. Other point of interest categories may typically be highly visible, for example restaurants. However, that does not mean automatically that the entities of this category stand out in their environment. For example, China Town may show a strip of Chinese restaurants. A POI service will unashamedly show the points of interest in their high local density. But just because of that reason no single restaurant stands out in its neighborhood. These restaurants are unsuited to form a point of reference in mental representations, and a decision point "At the China restaurant turn left" does not work either. The aggregate *China Town*, however, may form a landmark in some context.

This argument does not answer the opposite question either. Are landmarks points of interest? Some navigation systems, especially those addressing tourists, like to suggest so. In their context shown landmarks are recommendations to visit these places because they are famous, prominent, en vogue, of historical interest, or of cultural interest. If the purpose of the service is limited to recommendations only, neither the compilation of these landmarks for a POI database nor their presentation on a map aim to support human orientation. Nonetheless, whatever the purpose of the presentation these landmarks may actually help the tourist with global orientation after all. Accordingly, these landmarks are at most a subset of the landmarks studied in this book. They are those landmarks somebody has selected to make recommendations for tourists, i.e., those that serve also an interest different from spatial orientation and reasoning.

A special category of points of interest are those collected by a machine for an individual. Let us call them *favorite places*. Favorite places are a product of machine learning: They can be derived from the individual's prior search, movement behavior, their social network's behavior, or from the behavior of a group of people with similar profile. "My home" or "my favorite coffee place" can form landmarks in my mental spatial representation. These are individual landmarks. The challenge in communication is that these landmarks may not be shared, such that a place description "let us meet at my favorite coffee place" may not work, depending on the recipient's intimacy with my life. Landmarks imply a shared understanding, and thus, databases should store geographic objects that, since they can be experienced by all people, have chances to structure many people's mental representations.

1.2.3 Icons

Another term to distinguish landmarks from are icons. Icons have a strong semantics in semiotics. Their image stands for something else. If the icon, an image, stands for a geographic entity there are some parallels to landmarks. Consider, for example, the Eiffel Tower, which is an icon of Paris, if not of France. Typically being an icon of a geographic entity requires a containment relationship. The Eiffel Tower is an icon of Paris because it is *in* (or *part of*) Paris, even relatively central, in addition of being highly visible, standing out with a unique, unambiguous shape, and carrying a strong emotional attachment of locals and visitors alike. Hence, icons refer to geographic objects that are landmarks—the geographic objects stand out, are known, are referred to—but not every landmark has an iconic significance. For example, Federation Square in Melbourne is a landmark—probably every person in Melbourne knows it, and it is frequently used in route descriptions—but since it has no clear image it is not an icon.

1.2.4 Metaphorical Landmarks

It appears that landmarks are serving so well in structuring the spatial domain that the concept has been mapped into other domains successfully. One of these mappings happens when geographic reality—the physical world—is mapped to the metaphysical world. Then for example *heaven, paradise* or *cloud-cuckoo land* become landmarks—orientation points, places to be—which fulfil above's definition with the only exception that the space can no longer be experienced with human senses. Hence we do not consider landmarks in metaphysical space.

These mappings into other domains have been mentioned already. Merriam Webster's definition (Meaning 3) contained "very important events or achievements", which would be in the social domain and also in the temporal domain, with regard to providing structure in time. The *landmark victory*, a significant event in history, structures the memory of human experience both as a sudden shift of power relations, and with it into a before and after the victory.

1.3 Why Landmarks Challenge Intelligent Systems

The interaction between humans—their embodied experience, spatial cognition and spatial communication—and intelligent systems is the actual focus of this book. In this context landmarks are an object of interest because they support human decision making as intelligent systems should do. And yet, intelligent systems find special challenges in recognizing, contextualizing, and communicating landmarks. Let us have a look what these challenges are.

For that purpose let us first clarify what an *intelligent system* is, especially for the confusing notion of intelligence in its name. Given that a human is called intelligent we would like to see some of people's abilities in a machine that promise to support the human in decision making.

When Alan Turing wanted to decide the question: "Can machines think?" [46], he envisioned a machine that by its communication skills cannot be distinguished from a person. He suggested an imitation game that has become known as the Turing Test: If a human player cannot distinguish in a blind dialog whether the communication partner is human or machine then this partner has to be called intelligent. Turing's test has fascinated the field of artificial intelligence for so long because it argues by communication behavior, and does not require from the machine to understand and duplicate the ways humans sense and represent the world, or reason about the world. Sensing, representing and reasoning of a machine has made great progress since Turing's landmark paper, but is of course different from human ways. For example, robots are able to explore their environment, to learn about it, to form maps about it, and to orient themselves in this environment by landmarks. Yet each of these processes works internally quite different from the processes in the mind of a person. The robot sensors are different from human senses, the landmarks a robot finds useful are not necessarily the landmarks a person would identify and use in an environment, and the spatial representations are different. Now, the imitation game only requires that a person and a machine can communicate, or more precisely that a machine can communicate on the person's terms, independent of its internal structures.

Accordingly, John McCarthy explained more recently the term artificial intelligence in the following dialog[7]:

Q. What is artificial intelligence?
A. It is the science and engineering of making intelligent machines, especially intelligent computer programs. It is related to the similar task of using computers to understand human intelligence, but AI does not have to confine itself to methods that are biologically observable.
Q. Yes, but what is intelligence?
A. Intelligence is the computational part of the ability to achieve goals in the world. Varying kinds and degrees of intelligence occur in people, many animals and some machines.
Q. Isn't there a solid definition of intelligence that doesn't depend on relating it to human intelligence?
A. Not yet. The problem is that we cannot yet characterize in general what kinds of computational procedures we want to call intelligent. We understand some of the mechanisms of intelligence and not others.

Also, more recently critique of the imitation game has been expressed, for example by French [13, 14]. His critique is not about the validity of Turing's argument, but the relevance of the test. He argues that perfect imitation would include making the same mistakes. This is a critical argument for the domain of human spatial cognition, where it is well documented that people make *systematic* errors of judgement (some of them are discussed later in this book). Assuming a player of the imitation game has read about the weaknesses of human spatial judgement, she could try to take advantage of it in a Turing test and may triumph to identify the communication partner as a machine because it does not show these weaknesses. However, theoretically, the computer could know this as well and skew its (undistorted) results accordingly, even randomly, to perfectly mislead the player. But what would be the purpose of building such a machine? Wouldn't it be more useful, French asks, to leave the machine making undistorted results and communicate them in human terms to a 'user', which is a person in a concrete decision making situation? Wouldn't that be smart?

People do not only make systematic errors of judgement in their spatial reasoning, they are also varying in their spatial (communication) skills (e.g., [5]). Along the same line of argument, it does appear smart only to build machines of best (human) reasoning and communication skills. As French writes: "The way forward in AI does not lie in an attempt to flawlessly simulate human cognition, but in trying to design computers capable of developing their own abilities to understand the world and in interacting with these machines in a meaningful manner" ([14], p. 77). Instead, French postulates that, in order to achieve artificial intelligence,

---

[7]http://www-formal.stanford.edu/jmc/whatisai/whatisai.html, last visited 3/1/2014.

Can you tell me the way to the airport?	🔍

Fig. 1.7 Interpreting common language involves geospatial intelligence

computers are increasingly "capable of understanding, even if that understanding is not isomorphic to our own", and points to growing capabilities of computers in "[sensing,] representing and contextualizing patterns, making links to other patterns and analyzing these relationships" (*ibid*.1) that will shape our interactions with them.

Thinking ahead along these lines, we are reaching the age of *calm computing* [48] where 'the machine' is no longer a disembodied isolated apparatus (although IBM's Watson still was disconnected from the Internet when it won its game of *Jeopardy!* in 2011). Instead, with the Internet of Things, where every object is addressable, has sensors, can interact with the environment and communicate with other objects, 'the machine' gets embedded in our environment. Objects in the environment can sense our presence and our intentions, and hence determine, whether, for example, they can serve as landmarks in the current circumstances. They can communicate their readiness to our smart (shall we say: intelligent?) service that helps us navigate in this environment, or maintaining our orientation. The smart service can communicate back to the objects that they have been selected to support a particular task, and could be triggered to stand out even more as landmarks (e.g., turning on their lights) for a particular encounter or period of time.

In this context the original question of what is an intelligent machine poses itself now different from an imitation game. Nevertheless, a whole range of challenges has to be solved in order to build an intelligent system capable to communicate via landmarks. Among these challenges are understanding references to landmarks in verbal or graphical human place and route descriptions, understanding context of the conversation and relevance within this context, understanding personal preferences or knowledge of an environment, understanding the human embodied experience of environments, especially salience, and understanding to use landmarks effectively and efficiently in conveying information. Given the rest of the book will talk about approaches tackling them, let us go into depth here with the illustration of some of these challenges.

Consider a request a traveller might pose to an intelligent system: "Can you tell me the way to the airport?" (Fig. 1.7). Sounds simple? Surely, people would not have difficulties to answer this question in a manner that is easily understood and applied by the traveller (let's ignore here that some people are really bad in direction giving). They most likely would be able to resolve the meaning of the destination, which in this case is also a landmark, from context, and in their response they most likely would refer to some landmarks. They might say: "Sure. Tullamarine, you mean? Follow this street [points] to the hospital, then turn right. From there, just follow the signs", referring to a hospital as a landmark at a critical point of the route, namely the point where to turn right.

In some respect understanding and answering the request is not a too difficult question for an intelligent system either. An intelligent system is superior to humans in computing routes. It computes faster, processes more data, and produces more accurate results. An intelligent system can for example guarantee to compute the quickest route, and perhaps even include data about current traffic conditions, something our person in the example above cannot do. But to do so the system must know start and destination. If the intelligent system runs on the user's smart phone determining the start is possible using the sensors on board. But to determine the destination the system has to resolve this word 'airport', which is surprisingly tricky.

Obviously, 'airport' is ambiguous. The system knows many airports. Which one is the one the user has in mind? Choosing the most prominent one (as search engines are good at) would guide everybody to Atlanta, currently the world's largest airport. But this is probably not what the user had in mind. Choosing the one nearest to the user would be a better guess, but in the concrete example this would lead to a local airfield, not the international airport. Choosing the one most frequently visited by this person (another option for machine learning) would also be inappropriate. No, in this particular case it is the one where the inquirer wants to depart with a flight in a couple of hours from now. An intelligent system could determine that the function of an airport is to provide air travel, that air travellers need valid tickets bought in advance, and thus, could check whether this user has such a ticket to identify the airport. This is not only quite a complex reasoning chain, it requires also to assess all the various suggestions and reject the less likely ones.

After a route has been selected by the system, it has to communicate it in a way easy to understand and memorize by the user. Assume that the system is aware of the benefits of communicating by landmarks, then it has a challenge in selecting landmarks for this purpose. While a person needs not think twice when picking the 'hospital', an intelligent system knows thousands of objects along the route, of a variety of types and spatial granularities, from suburbs to ATMs, garbage bins and light poles. What is a good landmark? The one in the most outstanding color? A fire hydrant; not a good choice for a car driver, and too many of them along the route. The largest one? The best known one? Probably one of the objects at decision points, but which one? And is the landmark unambiguous ("the hospital") or ambiguous ("a hospital")? Is the landmark known to the inquirer, or at least recognizable in its identity, such that the system can use its name ("Royal Melbourne Hospital")? If so, does the inquirer also know how to find this landmark such that the first part of the route can be folded into a simple instruction: "You know how to find Royal Melbourne Hospital?" Klippel calls this spatial chunking [23]. We will come back to this later. Or can the inquirer at least recognize the type, as a hospital can typically be identified in its function from the outside by signs and certain functions at ground level? If not, should the landmark be described by its appearance: "A tall building on your right that is actually a hospital"? That is to say an intelligent system should integrate knowledge about the context of the enquiry, knowledge about the appearance of objects, analytical skills with respect to the route, and knowledge about the familiarity of the inquirer with the environment.

If we go back to the human-generated response, we may have now a better appreciation of the feat the human mind has accomplished in understanding the request. From the conversation context people guess that 'airport' most likely refers to Melbourne's international airport, Tullamarine, and they may open with a confirmative question ("Tullamarine, you mean?"). The response conforms to Grice's conversational maxims [20], especially to the maxims of relevance and of quantity ("Make your contribution as informative as required," and "Don't make your contribution more informative than is required"). It also applies some other remarkable principles. For example it mixes conversational modes between verbal description and pointing ("this street"), engaging with the user in an embodied manner. It also avoids to rely on quantities (such as in "after 493 m turn right") by using a qualitative and deliberately uncertain description ("at the hospital turn right"). A remaining uncertainty about distance is resolved by the reference to a landmark, the hospital. Finally, it folds all further instructions along the highway into one (spatial chunking), relying on knowledge in the world—the signage to the airport—thus avoiding more redundancy in the triangle between traveller, the environment, and the speaker.

1.4 Summary

Landmarks stand out in environments, and structure mental representations of environments in cognizing agents. They form anchors in mental spatial representations, markers, or reference points. They are essential for any spatial reasoning, for example, for orientation and wayfinding, and for any spatial communication. They appear in sketches, in descriptions of meeting points or routes. People use landmarks quite naturally.

This is in stark contrast to spatial information systems, such as car navigation systems or mobile location-based services. In their interaction with people they lack an ability to interpret people's references to landmarks, or to generate information by referring to landmarks. Their internal representations of spatial environments is certainly not based on landmarks, and thus, landmarks have to be brought in additionally to their internal representational structure, and properly integrated such that algorithms can add, revise, or select landmarks as needed in spatial dialogs.

Since landmarks are a concept of cognizing agents but not of systems, the challenge addressed in this book is to bridge the divide between the two. This book will catch up, suggest formal models to capture the concept of landmarks for a computer, and integrate landmarks in artificial intelligence. In order to inspire formal modelling on foundations of cognitive science the book has three main parts: cognitive foundations, computational models capturing landmarks, and communication models using landmarks for intelligent human-machine dialog.

References

1. Appleyard, D.: Why buildings are known. Environ. Behav. **1**(2), 131–156 (1969)
2. Aristotle: Physics. eBooks@Adelaide. The University of Adelaide, Adelaide (350BC)
3. Couclelis, H.: Aristotelian spatial dynamics in the age of GIS. In: Egenhofer, M.J., Golledge, R.G. (eds.) Spatial and Temporal Reasoning in Geographic Information Systems, pp. 109–118. Oxford University Press, New York (1998)
4. Couclelis, H., Golledge, R.G., Gale, N., Tobler, W.: Exploring the anchorpoint hypothesis of spatial cognition. J. Environ. Psychol. **7**(2), 99–122 (1987)
5. Daniel, M.P., Tom, A., Manghi, E., Denis, M.: Testing the value of route directions through navigational performance. Spatial Cognit. Comput. **3**(4), 269–289 (2003)
6. Dey, A.K.: Understanding and using context. Pers. Ubiquit. Comput. **5**(1), 4–7 (2001)
7. Dijkstra, E.W.: A note on two problems in connexion with graphs. Numer. Math. **1**, 269–271 (1959)
8. Downs, R.M., Stea, D.: Image and Environment. Aldine Publishing Company, Chicago (1973)
9. Fellbaum, C. (ed.): WordNet: An Electronic Lexical Database. The MIT Press, Cambridge (1998)
10. Frank, A.U.: The rationality of epistemology and the rationality of ontology. In: Smith, B., Broogard, B. (eds.) Rationality and Irrationality. Hölder-Pichler-Tempsky, Vienna (2000)
11. Frank, A.U., Raubal, M.: Formal specification of image schemata: a step towards interoperability in geographic information systems. Spatial Cognit. Comput. **1**(1), 67–101 (1999)
12. Freksa, C.: Qualitative spatial reasoning. In: Mark, D.M., Frank, A.U. (eds.) Cognitive and Linguistic Aspects of Geographic Space. NATO ASI Series D: Behavioural and Social Sciences, pp. 361–372. Kluwer, Dordrecht (1991)
13. French, R.M.: Subcognition and the limits of the turing test. Mind **99**(393), 53–65 (1990)
14. French, R.M.: Moving beyond the turing test. Comm. ACM **55**(12), 74–77 (2012)
15. Galton, A.: Fields and objects in space, time, and space-time. Spatial Cognit. Comput. **4**(1), 39–68 (2004)
16. Gärdenfors, P.: Conceptual Spaces. The MIT Press, Cambridge (2000)
17. Gärling, T., Böök, A., Lindberg, E.: Adults' memory representations of the spatial properties of their everyday physical environment. In: Cohen, R. (ed.) The Development of Spatial Cognition, pp. 141–184. Lawrence Erlbaum Associates, Hillsdale (1985)
18. Gigerenzer, G., Goldstein, D.G.: Reasoning the fast and frugal way: models of bounded rationality. Psychol. Rev. **103**(4), 650–669 (1996)
19. Goodchild, M.F.: Formalizing place in geographical information systems. In: Burton, L.M., Kemp, S.P., Leung, M.C., Matthews, S.A., Takeuchi, D.T. (eds.) Communities, Neighborhoods, and Health: Expanding the Boundaries of Place, pp. 21–35. Springer, New York (2011)
20. Grice, P.: Logic and conversation. Syntax Semantics **3**, 41–58 (1975)
21. Han, X., Byrne, P., Kahana, M.J., Becker, S.: When do objects become landmarks? A VR study of the effect of task relevance on spatial memory. PLoS ONE **7**(5), e35940 (2012)
22. Kahneman, D.: Thinking, Fast and Slow. Farrar, Straus and Giroux, New York (2011)
23. Klippel, A., Hansen, S., Richter, K.F., Winter, S.: Urban granularities: a data structure for cognitively ergonomic route directions. GeoInformatica **13**(2), 223–247 (2009)
24. Knauff, M.: Space to Reason: A Spatial Theory of Human Thought. MIT Press, Cambridge (2013)
25. Kuhn, W.: Modeling the semantics of geographic categories through conceptual integration. In: Egenhofer, M.J., Mark, D.M. (eds.) Geographic Information Science. Lecture Notes in Computer Science, vol. 2478, pp. 108–118. Springer, Berlin (2002)
26. Lakoff, G., Johnson, M.: Metaphors We Live By. The University of Chicago Press, Chicago (1980)
27. Lakoff, G.: Women, Fire, and Dangerous Things: What Categories Reveal about the Mind. The University of Chicago Press, Chicago (1987)
28. Lynch, K.: The Image of the City. The MIT Press, Cambridge (1960)

29. Mark, D.M., Smith, B., Tversky, B.: Ontology and geographic objects: an empirical study of cognitive categorization. In: Freksa, C., Mark, D.M. (eds.) Spatial Information Theory. Lecture Notes in Computer Science, vol. 1661, pp. 283–298. Springer, Berlin (1999)
30. Mark, D.M., Turk, A.G.: Landscape categories in Yindjibarndi. In: Kuhn, W., Worboys, M.F., Timpf, S. (eds.) Spatial Information Theory. Lecture Notes in Computer Science, vol. 2825, pp. 28–45. Springer, Berlin (2003)
31. Massey, D.: The conceptualization of place. In: Massey, D., Jess, P. (eds.) A Place in the World?, vol. 4, pp. 45–77. Oxford University Press, Oxford (1995)
32. Miller, G.A.: Wordnet: a lexical database for english. Comm. ACM **38**(11), 39–41 (1995)
33. Mohri, M., Rostamizadeh, A., Talwalkar, A.: Foundations of Machine Learning. MIT Press, Cambridge (2012)
34. Montello, D.R.: Scale and multiple psychologies of space. In: Frank, A.U., Campari, I. (eds.) Spatial Information Theory. Lecture Notes in Computer Science, vol. 716, pp. 312–321. Springer, Berlin (1993)
35. Presson, C.C., Montello, D.R.: Points of reference in spatial cognition: stalking the elusive landmark. Br. J. Dev. Psychol. **6**, 378–381 (1988)
36. Relph, E.C.: Place and Placelessness. Pion Ltd., London (1976)
37. Rosch, E.: Natural categories. Cognit. Psychol. **4**(3), 328–350 (1973)
38. Rosch, E.: Principles of categorization. In: Rosch, E., Lloyd, B.B. (eds.) Cognition and Categorization, pp. 27–48. Lawrence Erlbaum Associates, Hillsdale (1978)
39. Rosch, E., Mervis, C.B., Gray, W.D., Johnson, D.M., Boyes-Braem, P.: Basic objects in natural categories. Cognit. Psychol. **8**(3), 382–439 (1976)
40. Russell, S.J., Norvig, P.: Artificial Intelligence: A Modern Approach, 2nd edn. Pearson Education, London (2003)
41. Sadalla, E.K., Burroughs, J., Staplin, L.J.: Reference points in spatial cognition. J. Exp. Psychol. Hum. Learn. Memory **6**(5), 516–528 (1980)
42. Siegel, A.W., White, S.H.: The development of spatial representations of large-scale environments. In: Reese, H. (ed.) Advances in Child Development and Behaviour, pp. 9–55. Academic, New York (1975)
43. Sorrows, M.E., Hirtle, S.C.: The nature of landmarks for real and electronic spaces. In: Freksa, C., Mark, D.M. (eds.) Spatial Information Theory. Lecture Notes in Computer Science, vol. 1661, pp. 37–50. Springer, Berlin (1999)
44. Thrun, S., Burgard, W., Fox, D.: Probabilistic Robotics. MIT Press, Cambridge (2005)
45. Tuan, Y.F.: Space and Place: The Perspective of Experience. University of Minnesota Press, Minneapolis (1977)
46. Turing, A.M.: Computing machinery and intelligence. Mind **59**(236), 433–460 (1950)
47. Tversky, A.: Features of similarity. Psychol. Rev. **84**(4), 327–352 (1977)
48. Weiser, M., Brown, J.S.: The coming age of calm technology. In: Denning, P.J., Metcalfe, R.M. (eds.) Beyond Calculation: The Next Fifty Years of Computing. Springer, New York (1997)
49. Winter, S., Freksa, C.: Approaching the notion of place by contrast. J. Spatial Inform. Sci. **2012**(5), 31–50 (2012)
50. Wittgenstein, L.: Philosophical Investigations, 2nd edn. Basil Blackwell, Oxford (1963)

Chapter 2
Landmarks: A Thought Experiment

Abstract A thought experiment illustrates the fundamental role of landmarks for spatial abilities such as memory, orientation and wayfinding, and especially for human communication about space. We take a constructive approach, starting from a void environment and adding experiences supporting spatial abilities.

2.1 Experiment

In 1984 the neuroscientist Braitenberg presented a fascinating thought experiment by constructing increasingly complex 'vehicles' from a small set of primitive abilities [3]. Already with a few of these abilities combined the 'vehicles' showed a complex behavior hard to predict. Braitenberg introduced his experiment with the words ([3], p. 2):

> We will talk only about machines with very simple internal structures, too simple in fact to be interesting from the point of view of mechanical or electrical engineering. Interest arises, rather, when we look at these machines or vehicles as if they were animals in a natural environment. We will be tempted, then, to use psychological language in describing their behavior. And yet we know very well that there is nothing in these vehicles that we have not put in ourselves.

Braitenberg explained a complex system—such as the brain, or human behavior controlled by the brain—by demonstrating that a combination of a few simple modules already develops complex behavior. A similar approach of constructing complex systems from simple mechanisms was also taken by Couclelis [5] to explain self-organization of urban dynamics, or by Both et al. [2] to speculate how complex spatiotemporal behavior arises from spatiotemporal knowledge.

Let us devise an experiment in a similar vein. We pose a few simple modules for human orientation and wayfinding in an environment in order to see how the central role of landmark experiences arises quickly.

K.-F. Richter and S. Winter, *Landmarks: GIScience for Intelligent Services*,
DOI 10.1007/978-3-319-05732-3_2, © Springer International Publishing Switzerland 2014

2.1.1 The Void

Imagine an environment with no structure at all, like in Genesis 1.2: "The earth was without form, and void; and darkness was upon the face of the deep". The only structure given is a flat infinite surface orthogonal to gravity. Walkers are presented with a monochrome empty plane up to the horizon, under a white sky of diffuse light. There is no further structure in this environment, and no hint for direction other than the vertical axis imposed by gravity. There is not even shadow supporting a sense of direction. All what the walkers experience in this environment is their own locomotion, and thus path integration. Their body will tell them from which location they originated. Therefore they can always point in the direction of this location and guess the covered distance, a mental ability called homing [14]. In their desire to establish and maintain orientation in this empty environment the only location to relate to is this point of origin. Let us call it *home*. Home is the only place there is. Thus home becomes a reference point for the exploration of the environment: A landmark.

If this environment would have force fields that differ by location then sensing the force differentials could support the sense of direction. For example, if walkers would be equipped with a magnetic sense, or a compass as an external device, their mental effort to maintain their sense of direction would be supported considerably [21]. Similarly if the plane would be tilted towards gravity their sense of gravity would add observations to path integration.

2.1.2 Adding a Landmark

Now imagine that a walker, after roaming around to discover the environment, stumbles upon a coin on the ground. This walker, picking up the coin, might feel lucky. At last she has an experience in an otherwise uneventful environment. This experience is linked to a location. She will remember the event, and for a while also its location: Another landmark.

When she returns home she wants to report to her friends where she has found the coin. How can she do this given that there are no external cues in the environment? All she can refer to are directions and distances related to her body. She might say: "Over there [pointing], perhaps 20 steps from here", the direction physically linking to her body, and the distance in a quantity relating to an internalized measure that can be realized by walking. The reproducibility of measures proportional to the body or body mechanics will help her even in communicating with friends, assuming they have a body like her. Scheider et al. also concerned about grounding human experiences of space, wrote ([18], p. 76):

> Humans perceive length and direction of steps, because (in a literal sense) they are able to repeat steps of equal length and of equal direction. And thereby, we assume, they are able to observe and measure lengths of arbitrary things in this environment.

They then went on to develop a theory of steps between foci of attention. In our thought experiment we stick with the embodied experience, memory and communication of the walker (coming back to formal models in Chap. 4). In this regard it is interesting to see how instead of words the walker used pointing to communicate the direction. Finding words of similar accuracy in this environment would have been quite difficult. The distance, however, was expressed in a quantitative manner. A qualitative description might have come to mind more easily: "Over there [pointing], not too far from here". The qualitative description can be generated with less cognitive effort (we will come back to 'quality over quantity' in Sect. 3.3.2.1), but its realization might be more uncertain. In addition, the interpretation of a qualitative term is context-dependent; *not too far* does mean different things when talking about a car ride or an exploration of the immediate neighborhood, and in this empty environment there is not much shared experience between the walker and her friends that would establish context.

For her friends, the realization of her place descriptions is subject to uncertainty. While an instruction "20 steps" may produce less uncertainty than "not too far", it also takes more cognitive effort to realize by requiring counting. And since the coin was picked up by the walker, the landmark experience of her friends is only a mediated one.

Human bodies vary. For example, using foot lengths or step lengths as measures depends on an individual's body and can be reproduced by another person only with some uncertainty. Hence, when quantities need to be accurate some agreement is required on an absolute measure, which is a measure independent from an individual body. Most standardized unit measures are anthropomorphic, based either on average body dimensions (e.g., the length of a foot), or on human body movement (e.g., the length of a step). Even the *meter*, our today's standard unit according to the International System of Units (*SI*, from Système International d'Unités), was defined as a breakdown of the length of a great circle of the Earth to a unit in some relation to the scale of the human body. In order to be reproducible by everybody, an absolute distance measure requires from an individual learning their individual body properties (e.g., step length) compared to the standard. With other words, absolute measures require an additional layer of cognitive effort in realization.

Now, whether quantitative or qualitative, the tuple of distance and direction is well-known in geometry as polar coordinates. Coordinates are only meaningful within a given reference system, and in the given scenario the coordinates can only be embedded either in the egocentric reference system of the body of the speaker or of the recipient [11]. It is up to the speaker to convey the intended interpretation. For example, the speaker could have said: "Straight, not too far".

This expression could refer to the direction in front of herself ("put yourself in my position; straight in front of me"), or could have referred to the direction in front of the friend ("from your position walk straight"). For the prior interpretation, the recipient must transform the instruction by a mental rotation from the orientation of the speaker to their own orientation. For the latter interpretation, the speaker must do the mental rotation before speaking. Both is practical only when speaker and recipient are meeting face-to-face. If they are communicating over a distance (e.g., telephone) or asynchronously (e.g., email) the communication of body-pose related directions requires links to external cues. Since they do not exist in this environment such a communication is simply impossible [10]. Pointing, however, conveys the intended interpretation because it happens within the shared space [7].

What applies to the communication of direction—the ambiguity of the reference system, and the need for a mental transformation between reference systems either by the speaker or the recipient—applies also to the communication of distance. If the distance from the speaker is "about 20 steps" this information may need to be updated by the recipient according to their different positions and step sizes, and if the speaker actually means "about 20 steps in front of you" this interpretation needs to be conveyed as well. These mental transformations—rotations and translations—require spatial skills people have only to varying degrees [1, 19].

Stripped of any external cues within the environment, the walker will find it hard to describe accurately the location where she found the coin. The more time passes, or the more other walks she will have made since then, the less will she be able to reenact the locomotion experience. Constant updating of multiple vectors (home and all discoveries made over time) will become an overload, and the walker will give up maintaining those vectors felt no longer to be essential (last the homing vector). She may not necessarily forget the event itself, but she may forfeit her ability to describe its location. For a while though, the event provided a second landmark for the walker. In the real world we have similar experiences. "Let's meet at the café where we have met first" works in communication because this café has attached emotional value, is remembered for the meeting and for its location, and thus the location is describable and can be found again.

2.1.3 Adding Structure

From here on our thought experiment splits for a while into three parallel streams. One continues with constructing a memorable space (Sect. 2.1.3.1), another introduces a global frame of reference (Sect. 2.1.3.2), and a third one defines an arbitrary frame of reference (Sect. 2.1.3.3). Each of them ends up with a network structuring the environment, although motivated by different principles. The lines will be reunited in Sect. 2.1.3.4.

2.1.3.1 Adding More Landmarks

Let us assume the walker decides to mark the location where she found the coin with some chalk on the ground. She also draws a line on the ground from that location to home. Home is another mark on the ground. These externalized landmarks can now be found even with fading memories of path integration. The experience of walking can be repeated, can be communicated easier to others, and can be shared by others.

Further landmark experiences in the environment can be added over time, and connected to the existing ones. What forms over time is a travel network between landmarks. The intersections in this network were originally destinations, or locations of particular (shared) memories or stories. But over time also the edges in the network get some prominence, since they are commonly experienced by some embodied locomotion. As the dependent elements (e.g., "the route from home to the place where we have found the coin") their prominence may be lower, but we will argue later in the book that they will also share some landmarkness. For example, the walker and her friends can give names to edges (or sequences of edges in this regard). If over time the stories of the original landmark experiences fade away the prominence of the named edges may even get stronger (e.g., "the coin route").

Whatever elements are the primary anchors, either the connected landmark locations or the dual view of the edges between landmarks, this network enables for spatial tasks such as orientation and wayfinding. Using landmark orientation would be maintained either with local landmarks ("I am at the location where the coin was found") or with global landmarks ("I am three intersections from home"), and route planning would be about an appropriate sequence of landmarks (e.g., "From home to the location where the coin was found, and then right"). Using edge orientation would be maintained also either locally ("I am on Coin Route") or globally ("Coin Route must be in this direction"), and route planning would be about an appropriate sequence of edges (e.g., "Point Route, then turn right into Serendipity Street").

2.1.3.2 Adding Directed Light

Alternatively, let us change the settings of the experiment slightly. Instead of diffuse light imagine there is a point-like light source, mounted several times above body height. Call it *the sun* although it will not move in this experiment. This sun can be observed from any location, and since it is the only marked point in the environment—a singularity in an otherwise homogeneous empty space—it will attract attention from walkers. It also provides a reference direction for orientation and communication. Instead of using solely their locomotion-based orientation, walkers can now refer to the sun: "Walk towards the sun". Even on the ground is now a singular location where the body throws no shadow, which is where the sun is in zenith. This point can be found by any walker. It is an embodied experience but also a characteristic of the environment, and thus independent from previous locomotion. Therefore it can be used as a common, or shared reference point. Everybody can find it with no further instructions about its distance or direction.

Fig. 2.1 Similarity of
triangles between body height
(a), shadow length (b), height
of the light source (c), and
distance from the pole (d, to
the peak of the shadow)

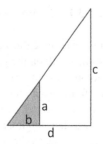

With respect to this reference point distances can be estimated: "Closer to the sun"
means a location where the walker has a shorter shadow (an embodied experience),
"near the sun" may have some context-dependent meaning related to shadow length,
and even quantities can be given, such as "within 10 step lengths from the pole under
the sun".

> Since in our thought experiment the height of the sun is related to human body
> dimensions, and constant, the walker and her friends could develop over time
> also a sense of distance from the pole by observing the angle of the sun above
> the horizon, or the length of their body's shadow, instead of estimating steps
> from path integration. With a constant height of the sun c and body height a,
> the shadow length b is proportional to the current distance from the pole $d - b$
> by the similarity of triangles (Fig. 2.1):
>
> $$\frac{a}{b} = \frac{c}{d}$$

Moreover, by applying projective geometry any walker can (re-)produce lateral
circles around the pole. In a salient distance from the pole (say, every 10 m) turning
orthogonally to their shadows they keep walking with keeping the sun constantly to
their right until they reach their starting point. These lines, imaginary or also drawn
by chalk, can play a role in the mental conceptualization of the space: "The coin was
above/below the lateral circle at 50 steps" may sound a bit arbitrary as an example,
but consider the decree by Pope Alexander VI that divided the world between Spain
and Portugal along a meridian of 100 leagues west of the Cape Verde Islands. This
line got a historic meaning, and had landmark character.

However, in order to relieve human memory from relying on tracked locomotion,
locations should be specified as points in space. We get there by adding only one
additional element to our experiment. Let us assume the walker, after finding the
coin, moves back straight to the pole (towards the light) and marks the direction to

Fig. 2.2 The Greenwich
meridian: an arbitrary choice
to anchor a geographic
coordinate system
(CC-BY-SA 3.0 *Rodney
Brooks*)

the location of the discovery by chalk on the ground. By doing so she defines a prime meridian (Fig. 2.2). Let us call the direction of the prime meridian *North*. With a marked direction and the memory for the distance the walker has now a reproducible characterization of the location of the discovery, one that does not change and is not dependent on her actual pose or location. Compared to path integration, requiring constant updating of all related locations, this is quite a relief for her memory. If she wants to find the location again she only has to come back to the pole (a landmark), find the prime meridian (another landmark), and memorize the distance. If she wants to tell friends she can now text: "Go to the pole, find the prime meridian, and walk about 20 steps", and neither of them has to be at the pole at the time of communication. Furthermore, future other discoveries can be linked to the pole and prime meridian as well. The pole becomes the datum point of a global reference coordinate system.

So, similarly to the world constructed from landmark experiences alone (Sect. 2.1.3.1), a global reference coordinate system relieves from constantly updating internal representations, and takes over to anchor other locations. Additionally we have gained a (polar) network structure by marking salient locations with reference to the datum—the pole and the prime meridian. The city of Karlsruhe in Germany, for example, shows such a radial network of lateral circles and meridians, laid out from the palace in the center (Fig. 2.3). Alternatively, the walker could lay out a rectangular network by constructing parallels to the prime meridian and then perpendiculars to the meridian. A rectangular network has the advantage of allowing constant block sizes, where the radial network has constantly increasing blocks with the distance from the pole. A rectangular grid is a street network pattern chosen in many European settlements in the new colonies of the eighteenth and nineteenth century, such as North America and Australia.

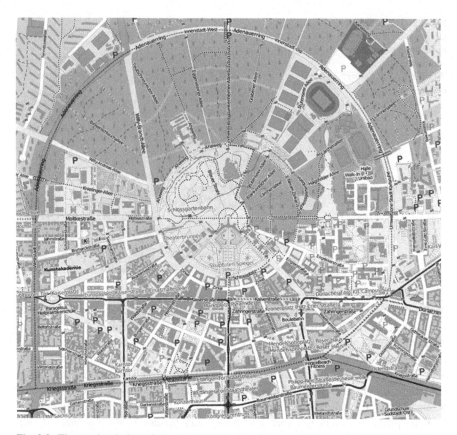

Fig. 2.3 The regular design of the city of Karlsruhe, Germany. Map copyright OpenStreetMap contributors, used under CC BY-SA 2.0

The walker decides to move from now on only along the drawn lines, and calls them *streets*. In contrast to Sect. 2.1.3.1 these streets are constructed from abstract principles, and so far only the datum has produced a landmark experience.

2.1.3.3 Adding an Arbitrary Network

In another alternative, let us assume the walker does not care whether the environment provides any cues for structure. Instead, the walker decides to draw freely a network of lines on the ground. The walker might be guided by cognitive efficiency, as too sparse lines on the ground would cause temptation to find shortcuts (adding to the network), and too dense lines would reduce the imageability of the network [12]. As in the other two lines of thought these streets would help structuring the environment, but would not be based on any prior landmark experiences. Most cities have neither a circular nor a rectangular network structure.

2.1.3.4 Networks Structuring an Environment

We are now ready to reunite the three alternative lines of our thought experiment. Whatever network the walker chooses, the result is a structure of nodes (say, street intersections) and edges between nodes (say, street segments between intersections). Such a structure is a graph [4, 9, 15, 22]. This graph, since it was drawn by the walker on the ground, is also planar (each intersection of edges is a node) and embedded (each node is at a particular location on the plane).

Each network structures the environment independent from prior landmark experiences. It has distinguished locations, the intersections, that are easy to perceive with the body senses as locations of choice. They are memorable, and hence landmarks by themselves. Since the walker could draw only a network of limited extent the individual intersections are countable and finite. Instead of polar coordinates within the polar reference system locations can now be described using this discrete network, e.g., "at the corner of". Distances can now be measured in numbers of intersections. Directions are discrete as well. In the radial and the rectangular network only right angles exist, and even in the free-form street networks intersections offer a very limited number of possible directions to take. In graph theory, this property is characterized by the *degree* of nodes: The degree of a node is the number of its incident edges. Radial and grid network have only nodes of degree 4, except the pole in the radial network and the outer boundary of the networks. Free-form street networks can also have nodes of degree 1 (dead-ends), degree 3 (e.g., T-intersections) and degrees higher than 4 (more complex intersections). However, due to physical space constraints this number cannot be arbitrarily large. From a cognitive perspective the limited number of directions is again a relief. The walker does no longer need to control constantly direction and to integrate steps. Instead, the walker only counts the passed intersections and memorizes discrete turn choices. These typically low counts are not stretching numerical cognition and short term memory [6, 13].

Networks may appear relatively plain, but then there are also individual differences coming out of node degrees or node (or edge) centrality. For example, the pole in a radial network has a node degree standing out in the otherwise regular structure, and it is also the node of highest betweenness centrality in the network— betweenness centrality of a node in a graph is a measure of how many shortest paths a node is in [8]. The latter means that statistically it will be experienced more often by walkers than other nodes. These reasons add to the pole's experiential features. The pole is a stronger landmark than the other intersections, and due to its uniqueness in this respect it is a *global* landmark. One function of a global landmark is supporting global orientation and wayfinding. For example, even if a walker somewhere in the radial network feels temporarily disoriented, some simple heuristics will lead her back to the pole. She will follow the next meridian, i.e., the straight streets towards the sun, and will reach the pole. In other network forms local variations may produce more subtle differences between nodes, but centers or bottlenecks will stand out as well. A regular node, however, is a *local* landmark. It helps locating events and referencing to these events as local anchor points.

For example, if the coin would have been found along an edge, not at a node, any of the following would be a natural reference: "Near corner West fourth St and Prime Avenue", or "In West fourth St, 30 steps from Prime Avenue", the latter implying an intersection of the two streets.

2.1.4   Filling with Further Landmarks

Now let us imagine the cells enclosed by network edges and nodes are filled with white blocks, larger than the human body. These blocks limit the sight of walkers. Since they are plain white without further texture or structure they are nearly indistinguishable for the human senses. There is no particular bodily experience attached to the encounter of any individual block except the shape of the cell they occupy.

These blocks get more importance in our experiment if we give them individual faces or meanings. Individual blocks can get attached a special shape, or a special color. A block can be labelled "supermarket", and another one "café". These blocks stand out from the other ones that were left unchanged plain white. They provide a special experience for passersby. Walkers will memorize these experiences, and attach them to the locations where they make these experiences. As long as these labelled objects are globally unique and only few they can have global landmark characteristics. "In the direction of the supermarket" provides global orientation if everybody knows the supermarket. They can also take the function of local landmarks. People can refer to it when describing local events ("in front of the supermarket", "three blocks from the supermarket"), either trusting that the recipient has made the encounter with the supermarket before already, or will easily identify it when passing by. "At the supermarket" is an even more efficient description than "at the third intersection" because it does require only one object recognition task, and no counting.

However, landmarks do not have to be globally unique. There might be a second supermarket in this environment, perhaps even of the same brand, as it happens out there in any real city. The two supermarkets are still differentiable from the rest of the environment, but an instruction such as "At the supermarket" must be considered ambiguous now. There are three common cognitive mechanisms that are used for disambiguation of local landmarks in a communication context:

1. *Nearness*: "At the supermarket" is disambiguated by choosing the nearest individual as a default. Behind this cognitive heuristics is also embodied experience since the cost of interaction with the environment is inversely proportional to the cost of travel: the nearest supermarket is the easiest one to reach. The argument is additionally supported by the first law of geography as stated by Tobler: "Everything is related to everything else, but near things are more related than distant things" ([20], p. 236).

2. *Order*: "Go straight and at the supermarket turn left" cannot be solved on the assumption that this *next* supermarket in a particular direction is also the nearest one. Instead a principle of order comes into play. The next supermarket is the first encountered in a particular search. By the same ordering principle one could equally well say: "At the second supermarket turn left".

3. *Hierarchical priming*: While "at the supermarket" is ambiguous, "at the supermarket on Prime Avenue" is less likely ambiguous (depending on whether the street name is a differentiator). This hierarchical localization [16,17] works recursively if needed, i.e., in cases where Prime Avenue either is not unambiguous or not prominent enough. "At the supermarket on Prime Avenue, in the Southern sector [of the environment]" is adding in this way. There is strong evidence that spatial mental representations are hierarchically organized.

Hierarchical priming happens also through salience hierarchies of landmarks. For example, a hierarchical description (or thought) "I found the coin in the entrance of the supermarket, not far from the ATM" refers to the supermarket as a global landmark, and then specializes further by another reference to a local landmark, an ATM. Here the hierarchical priming is required for disambiguation between the many ATMs in an environment. This way, hierarchies help to break down large environments into manageable regions of influence. Any object or event in the environment can be related to landmarks in a variety of ways, qualitatively and quantitatively. Besides of something being "near the supermarket" (distance, defining the region of influence), its location can also be characterized as "on the way from supermarket to home" (orientation), as "on the right when travelling from supermarket to home" (direction), as "between the supermarket and home" (projection), or as "in the quarter of the supermarket" (topology). Quantitative characterizations are possible as well, such as "30 m from the supermarket". Also the prominence of a landmark can prime the memory for a particular street segment, as in "I found a coin in the street where the supermarket is".

2.2 Summary

A thought experiment has illustrated the fundamental role of landmarks for structuring mental representations of an environment. In the constructive approach of the experiment we have assumed an environment that provides salient experiences linked to locations, and expected that these experiences form anchor points in memory, and also configurations that allow relative orientation. We have in particular learned that landmark configurations are sufficient for spatial cognitive tasks such as orientation and wayfinding. The existence of a global frame of reference, which is essential in any technical system, from spatial information systems to robots, is not essential for human spatial problem solving as long as the environment provides configurations of landmarks allowing relative orientation. We have also learned that experiences at particular locations form landmarks, and these experiences can also be experiences of the structure (nodes) or dimensions (edges) of the structure of an environment.

The fact that our environment resembled an urban environment does not matter. The same insights about landmarks could have been made in a process introducing landmarks in a landscape. Spatial cognition certainly developed in natural environments, but its mechanisms apply equally in human-made environments.

Landmarks' primary role appears to be structuring a mental representation of an environment by forming anchors for relational links. Verbalized, these anchors and their links appear to convert into relational descriptions. Evidence for this assumption will be presented in the next chapter. Without landmarks both tasks, forming a mental representation of an environment and, correspondingly, being able to communicate about locations in the environment, appear to be significantly more complex, relying on locomotion and path integration only.

References

1. Battista, C., Peters, M.: Ecological aspects of mental rotation around the vertical and horizontal axis. J. Indiv. Differ. **31**(2), 110–113 (2010)
2. Both, A., Duckham, M., Kuhn, W.: Spatiotemporal Braitenberg vehicles. In: Krumm, J., Kröger, P., Widmayer, P. (eds.) 21st ACM SIGSPATIAL International Conference on Advances in Geographic Information Systems. ACM Press, Orlando (2013)
3. Braitenberg, V.: Vehicles: Experiments in Synthetic Psychology. The MIT Press, Cambridge (1984)
4. Christofides, N.: Graph Theory: An Algorithmic Approach. Academic, London (1975)
5. Couclelis, H.: Of mice and men: what rodent populations can teach us about complex spatial dynamics. Environ. Plann. A **20**(1), 99–109 (1988)
6. Dehaene, S.: The Number Sense. Oxford University Press, New York (1997)
7. Emmorey, K., Reilly, J.S. (eds.): Language, Gesture, and Space. Lawrence Erlbaum Associates, Inc., Hillsdale (1995)
8. Freeman, L.C.: A set of measures of centrality based on betweenness. Sociometry **40**(1), 35–41 (1977)
9. Harary, F.: Graph Theory. Addison-Wesley, Reading (1969)
10. Janelle, D.G.: Impact of information technologies. In: Hanson, S., Giuliano, G. (eds.) The Geography of Urban Transportation, pp. 86–112. Guilford Press, New York (2004)
11. Klatzky, R.L., Loomis, J.M., Beall, A.C., Chance, S.S., Golledge, R.G.: Spatial updating of self-position and orientation during real, imagined, and virtual locomotion. Psychol. Sci. **9**(4), 293–298 (1998)
12. Lynch, K.: The Image of the City. The MIT Press, Cambridge (1960)
13. Miller, G.A.: The magical number seven, plus or minus two: some limits on our capacity for processing information. Psychol. Rev. **63**, 81–97 (1956)
14. Mittelstaedt, M.L., Mittelstaedt, H.: Homing by path integration in a mammal. Naturwissenschaften **67**(11), 566–567 (1980)
15. Ore, Ø.: Graphs and their uses. In: New Mathematical Library, vol. 34. The Mathematical Association of America, Washington (1990)
16. Plumert, J.M., Spalding, T.L., Nichols-Whitehead, P.: Preferences for ascending and descending hierarchical organization in spatial communication. Mem. Cogn. **29**(2), 274–284 (2001)
17. Richter, D., Vasardani, M., Stirling, L., Richter, K.F., Winter, S.: Zooming in–zooming out: hierarchies in place descriptions. In: Krisp, J.M. (ed.) Progress in Location-Based Services, Lecture Notes in Geoinformation and Cartography. Springer, Berlin (2013)

18. Scheider, S., Janowicz, K., Kuhn, W.: Grounding geographic categories in the meaningful environment. In: Stewart Hornsby, K., Claramunt, C., Denis, M., Ligozat, G. (eds.) Spatial Information Theory. Lecture Notes in Computer Science, vol. 5756, pp. 69–87. Springer, Berlin (2009)
19. Shepard, R.N., Metzler, J.: Mental rotation of three-dimensional objects. Science **171**(3972), 701–703 (1971)
20. Tobler, W.: A computer movie simulating urban growth in the Detroit region. Econ. Geogr. **46**(2), 234–240 (1970)
21. Turner, C.H.: The Homing of Ants: An Experimental Study of Ant Behavior. University of Chicago, Chicago (1907)
22. Wilson, R.J., Watkins, J.J.: Graphs: An Introductory Approach. Wiley, New York (1990)

Chapter 3
Cognitive Aspects: How People Perceive, Memorize, Think and Talk About Landmarks

Abstract This chapter deals with the human mind and its representation of geographic space, particularly with the role of landmarks in these representations. The scientific disciplines that are called upon to illuminate this area are neuroscience, cognitive science, and linguistics. This broad range of disciplines is necessary, because the structure of spatial representations in the human brain and the behaviour of these representations in spatial tasks are not directly accessible, and thus, indirect approaches have to be pursued. Direct observations of brain cells are invasive, and thus applied typically on animals only. To what extent observations from animals can be transferred to explain human spatial cognition is a matter for investigation in its own right. However, indirect methods such as functional magnetic resonance imaging shed some light into brain activity. Cognitive scientists, being interested in intelligence and behaviour rather than actual cell structures, live with a similar challenge. They observe the human mind indirectly by devising experiments on human memory, reasoning, and behaviour. Linguists add studies of human spatial communication, which should also allow indirect conclusions about mental representations.

3.1 Spatial Cognition

In his classic *Frames of Reference* [60] Gardner studied the specializations of abilities in the human mind. Considering the very different talents of, say, the religious scholar, the concert piano soloist, or the biomolecular scientist, it is clear that a standardized test of general intelligence cannot do justice to the variety in which human intelligence can take form. In order to be able to isolate specialized 'intelligences' Gardner set up criteria, namely that a specialization "must entail a set of skills of problem-solving [...] and must also entail the potential for finding or creating problems" (p. 64f.). He listed also additional evidence ("signs") for specializations, such as a potential isolation by brain damage, the existence of exceptionally performing individuals, or an identified set of core operations.

K.-F. Richter and S. Winter, *Landmarks: GIScience for Intelligent Services*, 41
DOI 10.1007/978-3-319-05732-3_3, © Springer International Publishing Switzerland 2014

The first sign, a potential isolation by brain damage, assumes that particular areas in the brain are responsible for particular tasks (or abilities). While these associations are generally observable for some spatial tasks—see for example [181]—looking closer into spatial abilities this argument might be difficult to prove. Many spatial abilities combine activities in different brain regions, not to mention the plasticity of the brain. One reason may be that spatial skills are themselves quite diverse, and hence not necessarily located in a single brain region. Take for example the range between visual-spatial abilities (e.g., object recognition), spatial memory abilities (e.g., object localization), or the cognitive elements of sensorimotor abilities (e.g., self-localization). In their own way, however, each of these abilities is interacting with landmarks.

Despite the breadth of spatial abilities, applying these criteria and further evidence, Gardner [60] identified a small set of 'intelligences', and among these a *spatial intelligence*. In some sense spatial abilities are fulfilling the other two signs. They are problem-solving skills, for example, recognizing or imagining an object from an unfamiliar angle, coordinating body movement or solving more complex tasks like wayfinding. Spatial abilities support everyday activities, such as finding home again, but also professional skills such as playing the piano (sensorimotor), imagining a DNA sequence (visual-spatial), or navigating an airplane (a combination). Some people feel good at spatial intelligence, and others will have difficulties with, for example, reading a map, or explaining a route. Central to Gardner's argument is that these abilities cannot be fully replaced (or explained) by other capacities, for example verbal or logical-mathematical intelligence. Others have even pointed out that spatial abilities help reasoning in non-spatial domains [154]. Nevertheless, spatial intelligence remains built on a variety of abilities. These spatial abilities can even be differently pronounced in individuals. For example, some may be acute in visual perception, but bad in self-localization. The individual performance in these skills can be tested, and these tests are quite popular.

In this book we do not use the term *spatial intelligence*, but its more established synonym *spatial cognition*. However, while the composite of spatial abilities define spatial cognition, confusingly the scientific study of these abilities is also called spatial cognition. The *Handbook of Spatial Cognition*, for example, states: "Spatial cognition is a branch of cognitive science" [231]. In effect it should be called spatial cognitive science. Anyway, relating back to the superordinate category, *cognitive science*, may not immediately help clarifying when eminent researchers say:

> Cognitive science is not yet established as a mature science [...] it is really more of a loose affiliation of disciplines than a discipline of its own. Interestingly, an important pole is occupied by Artificial Intelligence [...] other affiliated disciplines are generally taken to consist of linguistics, neuroscience, psychology, sometimes anthropology, and the philosophy of mind ([227], p. 4).

Accordingly, spatial cognitive science has many flavors or shades. It "seeks to understand how humans and other animals perceive, interpret, mentally represent and interact with the spatial characteristics of their environment" [231]. "The study of knowledge and beliefs about spatial properties of objects and events in the world" [148] can be approached from a neuroscience perspective (which regions in the brain, and which cell types, store and process spatial knowledge and are active in the processing of visual or sensorimotor stimulations), from a psychological perspective (how do people behave in orientation and wayfinding tasks, and what do their abilities reveal about cognitive capacities), from a linguistic perspective (how people communicate (and hence think) about space), from an anthropological perspective (for example, why Inuit are able to find home in their monotonous environment[1]), from a philosophy of mind perspective (for example, recognizing that landmarks form a graded category), and, last but not least, from the perspective of artificial intelligence. Artificial intelligence has two interests in spatial cognitive science, aligned with what Searle has called strong AI and weak AI [193]. One searches for computational models of human spatial cognitive abilities (imitating human thinking), the other one searches for the spatially intelligent machine, which is a machine able of interacting with humans on spatial problem solving (simulating human thinking).

Spatial cognitive science supports strongly our position that landmarks are embodied experiences that shape mental spatial representations. Montello states:

> Cognition is about knowledge: its acquisition, storage and retrieval, manipulation, and use by humans, non-human animals, and intelligent machines. Broadly construed, cognitive systems include sensation and perception, thinking, imagery, memory, learning, language, reasoning, and problem-solving. In humans, cognitive structures and processes are part of the mind, which emerges from a brain and nervous system inside of a body that exists in a social and physical world. Spatial properties include location, size, distance, direction, separation and connection, shape, pattern, and movement ([148], p. 14771).

and continues, focusing on the interrelationship between embodied experience and spatial cognition:

> Humans acquire spatial knowledge and beliefs directly via sensorimotor systems that operate as they move about the world. People also acquire spatial knowledge indirectly via static and dynamic symbolic media such as maps and images, 3-D models, and language. [...] Spatial knowledge changes over time, through processes of learning and development. [...] A person's activity space—the set of paths, places, and regions traveled on a regular basis—is an important example of spatial experience that influences people's knowledge of space and place, no matter what their age. Most people know the areas around their homes or work places most thoroughly, for example ([148], p. 14772).

Similarly, Varela et al. argue that cognition is embodied: "[We] emphasize the growing conviction that cognition is not the representation of a pregiven world by a pregiven mind but is rather the enactment of a world and a mind on the basis of a history of the variety of actions that a being in the world performs" ([227], p. 9), a view that is shared also by others [113].

---

[1]Inuit demonstrate significantly higher levels of visual memory [101].

Broader introductions to spatial cognition, both the spatial abilities as well as the methods of studying these abilities, are provided in a number of books, for example [28, 43, 68, 95, 120, 160, 167, 231].

3.1.1 Spatial Abilities

Spatial cognition is the ability of living beings to perceive, memorize, utilize and convey properties about their spatial environment. According to what we just have said, we should rather speak of a combination of abilities. But factorizing these abilities has been proven tricky. McGee, for example, identified an ability for *spatial visualization*—the ability to mentally manipulate, rotate or twist objects— and an ability for *spatial orientation*—the ability to imagine an object from different perspectives [140]. These two abilities were already part of the broader Guilford– Zimmerman aptitude survey ([73], parts V and VI). Later, Carroll identified five major spatial abilities, differently factorized, but in essence adding dynamic spatial abilities of estimating speed and predicting movements [21]. While these abilities could be identified and tested in paper-and-pencil tests, i.e., in small-scale space, finding environmental spatial cognitive abilities such as wayfinding or learning the layout of environments requires field experiments in environmental space [80, 198]. As environmental spatial abilities are also amalgams, Allen undertook an attempt to break down the spatial abilities that service wayfinding [3, 5]. He especially distinguished between a family of abilities dedicated to object identification, object localization and traveler orientation, and made finer distinctions within these families depending on whether the objects are static or mobile, and whether the observers are stationary or moving observers.

A recent review of spatial abilities and their individual performance measurements has been provided by Hegarty and Waller [78]. Individual performance variations in spatial abilities have triggered questions whether these abilities can be strengthened by training, and whether they are gender dependent.

The issue of gender and spatial abilities is hotly debated in colloquial contexts, but also in science. See for example Silverman and Eals' theory that sex differences exist and are grounded in evolutionary division of labor [199], or Dabbs et al. studying the frequently reported advantage of female in landmark-based and egocentric orientation and male advantage in cardinal/Euclidean orientation [34]. However, without going into detail here, across a large range of contributions the reported research results are contradictory. More fundamentally, the underlying assumptions are questionable since experiments do usually not distinguish between genetic disposition and socialization. With the high plasticity of the brain it may even be impossible to resolve this latter point in principle.

Fig. 3.1 Folding this sheet of paper along the *dotted line*, what will the result look like?

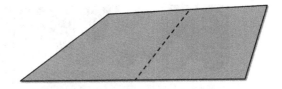

Refraining from siding with any particular factorization scheme, let us highlight just a few spatial abilities which we will relate to later.

3.1.1.1 Visualizing Objects

People find it easy to visually imagine intimately familiar spaces, even without visual stimulation. For example, people typically can answer a question such as "Imagine entering your living room—now what's to your left?" They even can manipulate these imaginations. Figure 3.1 shows a (picture of a) sheet of paper with a dotted line printed in the center. A test for the ability of mentally manipulating this object is: "Imagine folding it and viewing it from another angle. What will it look like?"

3.1.1.2 Relating Objects

Another visual-spatial ability helps understanding the spatial relationships of objects to each other [73], including object recognition. It develops early in childhood and is linked also to the emergence of stereo-(depth-)vision. People understand distances between objects and proportions of objects, which are essential skills for tasks such as packing a suitcase, constructing efficient mental models of objects from verbal or pictorial descriptions (Fig. 3.2), or interpreting a map.

3.1.1.3 Mental Rotation

People can form mental three-dimensional images of depicted objects, and rotate these images mentally. In their famous experiment Shepard and Metzler asked participants whether two figures are the depictions of the same object (Fig. 3.3). The time they take for deciding is proportional to the angular rotational difference, which is evidence for an actual mental rotation [197]. Based on Shepard and Metzler's stimuli, Vandenberg and Kuse developed later a pencil-and-paper test that has become a widely accepted standard test for mental rotation [225]. A prominent application of mental rotation is in map reading, where conventionally maps are oriented north-up, independent from the current orientation of the map reader. Wall-mounted maps are an obvious case for this argument since they need also to be mentally rotated to the horizontal plane [150]. Mental rotation must also be at

Fig. 3.2   In order to interpret the picture visual-spatial abilities use perspective and experience

Fig. 3.3   One of Shepard and
Metzler's mental rotation
tasks: are the two objects
identical?

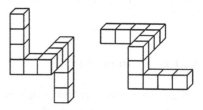

work when people take (or change) perspectives. Tversky and Hard, for example, have shown that people switch with ease between egocentric and another person's perspectives in their verbal descriptions [222].

3.1.1.4   Path Integration

People are able to locate themselves during locomotion automatically and continuously, even without landmarks as external references [50]. For example, this ability enables people to always point in the direction to the location they started from, independent from the route taken. Pointing gestures and tracking of path completion have been applied in studying human path integration capacity [125, 126]. Path integration is an ability essential for survival, not only for humans. Desert mice, for example, have a capability for homing [145]. Desert ants travel some tortuous routes outbound from home foraging for food, but then return home along straight routes [152, 234], proving to be able for path integration. Honeybees communicate directions and distances of food sources, information also gained from path integration [58]. For people, it guarantees finding home again even in the dark, or with a visual impairment. Since path integration works without external reference points, it has been linked to the sensor-motor system and vestibular organs, and in the brain to particular cell types in different regions of the brain. Path integration is closely related to our *sense of place* and *sense of direction* [79]. Path integration is accumulating uncertainty, and hence typically combinations with spatial updating are applied.

3.1.1.5 Spatial Updating

In order to be able to act, people know the position of surrounding objects relative to their body. They are spatially aware, and this awareness is provided by their senses. Since people act in space and time, their own pose or location changes constantly. Similarly objects in the environment can change pose or location. Only continuous updating of the spatial mental representation of these relationships helps with survival. Wolbers et al. studied spatial updating from a neuroscience perspective, i.e, how the brain keeps track [240], and Kelly et al. looked at spatial updating from a cognitive perspective, i.e., which cues in the environment are used [97, 98].

Hegarty et al. published a self-reporting measure—by introspection and self-assessment—of spatial abilities, called the Santa Barbara Sense of Direction Scale [79]. The questionnaire consists of several statements about the participant's spatial and navigational abilities, preferences, and experiences. Participants have to rank themselves on a seven-point Likert scale [123]. The self-reporting measure proved to be internally consistent and had good test–retest reliability, but of course does not provide an evaluation of actual performance.

1. I am very good at giving directions.
2. I have a poor memory for where I left things.
3. I am very good at judging distances.
4. My "sense of direction" is very good.
5. I tend to think of my environment in terms of cardinal directions (N, S, E, W).
6. I very easily get lost in a new city.
7. I enjoy reading maps.
8. I have trouble understanding directions.
9. I am very good at reading maps.
10. I don't remember routes very well while riding as a passenger in a car.
11. I don't enjoy giving directions.
12. It's not important to me to know where I am.
13. I usually let someone else do the navigational planning for long trips.
14. I can usually remember a new route after I have traveled it only once.
15. I don't have a very good "mental map" of my environment.

In other experiments with human participants, spatial coordination can be tested, such as catching a ball or predicting where a ball that is temporarily obscured by an object will reappear; orientation tasks can involve maintaining a sense for a reference direction during locomotion; wayfinding tasks can involve realizing a symbolic or verbal route description or exploring an unknown environment; and spatial communication tasks can involve providing such route descriptions.

3.1.2 Orientation and Wayfinding

How to get from here to there is certainly a fundamental planning ability of any animal, humans included. Montello and Raubal ranked wayfinding as the most important function of spatial cognition [151]. Aspects of wayfinding have already been mentioned, such as the homing ability based on path integration. Let us search for a more systematic understanding of this task of spatial cognition.

Montello [149] identified wayfinding as a component of *navigation*. He defines navigation as goal-directed movement of one's self through an environment to find a distal destination. However, splitting navigation into components does not happen along clear lines. While Montello proposes to split into *locomotion* and *wayfinding* (p. 258), Waller and Nadel see wayfinding as agglomerate of a suite of cognitive abilities such as place memory, imagery and planning [231]. Wang and Spelke [233] suggest to isolate a third component in between locomotion and wayfinding, *spatial orientation*. Let us take a closer look at these three components of navigation.

In this book we apply Montello's notion of wayfinding, which has also been adopted elsewhere (e.g., [12, 61, 66, 77]). It involves several abilities:

> *Wayfinding* comprises the tactical and strategic part of solving the problem to find a distal destination.

Allen [3] attempted a classification of types of wayfinding tasks. He distinguished wayfinding through familiar environments (he calls this type *commute*), through or into unfamiliar environments for the purpose of learning the environment (*explore*), and through or into unfamiliar environments to reach a destination known to exist (*quest*). Wayfinding of a commute type accesses an existing mental spatial representation. Exploring, in contrast, will contribute to a mental spatial representation. For questing, as Allen writes, "the only way in which a previously unknown place becomes a wayfinding destination is for knowledge of that place to be conveyed symbolically, but means of either a map or verbal descriptions" (p. 555). To what extent this questing contributes to mental spatial representations seems to vary in different contexts. Recent studies indicate that car drivers relying on their car navigation system put less effort into adding to their mental spatial representations [89]. Also the degree of detail in maps has an impact, as more abstract representations support the construction of a mental spatial representation better than, say, satellite imagery [162, 186].

Spatial orientation comes interlinked with locomotion (as spatial updating), utilizing the sensorimotor system, but also comes interlinked with wayfinding out of cognitive, sometimes even conscious effort. Thus spatial orientation involves also a bundle of spatial abilities. While spatial orientation may focus more on the basic ability of imagery, wayfinding may focus more on the basic ability of planning.

> *Spatial orientation* is the ability to create or maintain an image from sensory perceptions of the position of the body relative to the environment.

Maintaining the spatial orientation of the body to the outside world is a crucial condition for wayfinding and locomotion. Gärling et al. [62] differentiate several levels of orientation functions:

> We define environmental orientation as the ability to perceive one's position relative to points or systems of reference in the environment. These points or systems of reference may be perceptually available, but, [...] this is not a necessary condition. [...] A hierarchy of orientation functions may be tentatively proposed. Body orientation is defined as the perception of the body axes relative to the line of gravity and the limbs relative to the body axes. At the next level, orientation is maintained in the environment relative to perceptually available reference systems. Visual direction and position constancy as well as auditory localization are important here. At the highest level of the hierarchy, orientation is maintained relative to points and systems of reference that may not be perceptually available. Geographic orientation could be considered a special case. Other indirect information sources, such as maps, signs, and the sun, are available (p. 165).

The sensorimotor component of this task is controlled by the nervous system. The immediate orientation of the body is observed by the visual and auditory senses, head orientation and motion is registered by the vestibular (equilibrium) organ, and an awareness of the position of the parts of the body and their movements is maintained by the proprioceptive sense. An integration of these sensory inputs allows for coordinated actions of the body, for example, tracking targets, or controlling of posture, gait and other movements.

The cognitive component of this task requires a mental spatial representation of the environment. Cognition is constantly trying to maintain a subjective sense of orientation. This sense of orientation might be satisfied with the ability to locate oneself in one's mental representation of the environment. This explanation would be roughly equivalent to orientation of robots, establishing or maintaining a correspondence between sensor observations and the robot's internal representation of the world. For people, however, this definition is problematic since human sensing is subjective and biased by attention. Furthermore, their mental spatial representation is not directly accessible for observation, and neither are the individual's self-localization in this representation, the correctness of the self-localization, or the individual's degree of feeling oriented. In order to reduce the amount of subjectivity in the status of being oriented an individual should be able to externalize their current relationships with their environment. An indirect indicator, however, is their success with planning of goal-directed behavior in space.

The third component of navigation, locomotion, can be defined as follows:

> *Locomotion* is the actual movement of the body coordinated to the proximal surrounds.

The proximal surrounds can be described as the "environment directly accessible to our sensory and motor systems at a given moment (or, at most, within a few moments)" ([149], p. 259). Not all locomotion is part of an effort to find a distal destination. Locomotion can happen in vista space, for example, moving from the chair to the window, sitting at the table and leading the fork to the mouth, turning to another person, or walking down the aisle. But in the context of navigation locomotion is guided by a conscious plan, the wayfinding result.

Psychological experiments on wayfinding tend to focus on walking as the means of movement. However, in the context of navigation the movement of the body can be aided by vehicles. Riding a bike or a car still requires movement coordinated to the proximal surrounds in a directed manner. Some of these aids, however, do not require body movement at all. Sitting in a taxi or on a bus can be part of the strategy to find a distal destination, but the only body movement involved is getting on or off the vehicle.

Locomotion, orientation and wayfinding are inherently linked. While locomotion follows a plan made by wayfinding, in return, orientation and wayfinding are constantly informed by locomotion, or by the sensory perceptions during changing of place or pose, and updated accordingly. For example, while sitting on a bus, watching time and progress of the trip—cognitive activities supported by perceptions—the decision to get off according to plan (or even updating a prior plan) is triggering locomotion.

3.1.3 What We Have Learned About Landmarks from Orientation and Wayfinding

The fundamental role of landmarks for orientation and wayfinding stems from a strong correspondence between an experience captured in (spatial) memory and a location in the physical environment. While we have defined landmarks as the reference points of mental spatial representations, their corresponding physical objects can be called landmarks only with regard to their potential to produce an experience that will be captured in spatial memory.

Siegel and White have looked at the development of a mental spatial representation of an environmental space by somebody unfamiliar to this environment [198]. Siegel and White studied the first-hand experience of the environment by locomotion. However, environments can also be learned from secondary sources such as maps, sketches, verbal route descriptions or tourist guides. Thus, the following sequence of learning, described by Siegel and White, is not a step-by-step process but rather a complex continuous process [147].

During locomotion, any outstanding experience along the route will trigger a memory, located in space roughly by path integration. Siegel and White call this type of knowledge in the emerging mental spatial representation *landmark knowledge*. Landmark knowledge can also be mediated. The armchair traveller to Paris, reading about the highlights of Paris in a tourist guide and locating them on a map, will have a similar experience.

The path integration between landmark experiences provides connections called by Siegel and White *route knowledge* and elsewhere also *procedural knowledge*. The mode of locomotion, effort, and intensity of experience all impact on the experience of distance. Contributing to the sense of distance are motor sense, visual and auditory sense, but also the memory loading along the route, that means the number of objects or events experienced along the route, the degree of unfamiliarity with an environment, the mental preoccupation, and many more factors. Those who learn from secondary sources infer route knowledge from reading a tour description or reading a map [69].

A third tier of knowledge, called *survey knowledge* by Siegel and White, and also called *configurational knowledge*, emerges from integrating routes over time into a network-like representation that can be analyzed for directions and distances. Route segments can be freely recombined for wayfinding and orientation. However, people seem to have different cognitive preferences for representing an environment, or short, cognitive styles [163]. They are more landmark-focused, route-focused, or survey-focused. Their cognitive styles are correlated with spatial abilities [164]. According to Ishikawa and Montello some people even do not develop a survey representation, irrespective of the frequency of exposure to an environment or the activity in the environment [87].

Differing classifications were used as well. Piaget and Inhelder argued for a *figurative knowledge*—visual imagery of objects and configuration of objects— and *operative knowledge*—the ability to manipulate the visual imagery [166]. More recently, Gardner has suggested to make a distinction between relatively static and relatively active forms of spatial knowledge [60]. *Declarative knowledge* has been contrasted with *procedural* knowledge by Mandler [131]. Since declarative spatial knowledge lends itself to visual imagination it is similar to figurative or configurational knowledge.

Whichever classification is used, landmarks form the glue. Equipped with a representation of such knowledge orientation and wayfinding become possible. For example, Allen [3] has proposed a framework for examining the cognitive abilities involved in wayfinding. The framework consists of wayfinding tasks on one hand and means to accomplish these tasks on the other hand. In all of his identified tasks either landmark monitoring (e.g., for piloting) or landmark-movement memory (e.g., for path integration) are required. Steck and Mallot studied preferences for local or distant landmarks in visual navigation (piloting) [204]. Their results indicate that people use both in navigation, but individuals show preferences for one or the other, or use different types in particular locations. Golledge, studying human wayfinding behavior [67], observed the following route choice criteria most frequently: *shortest route, quickest route, route of fewest turns, most aesthetic route, first noticed route*, and *longest-leg-first route*. Perhaps not surprisingly, all of these criteria are linked to landmarks. Further evidence shows that distance estimates (in space or time) are correlated with the number of landmarks encountered along the route [182, 183, 214]. Accordingly, route choice can be distorted. The route of fewest turns is the one with the least cognitive load of landmarks in short term memory. The most aesthetic route could be the one passing a number of landmarks. The first noticed route may stand out as leading along the landmarks of strongest salience. The longest-leg-first route starts out with a single landmark ahead, such that more details can be loaded to short term memory later.

Siegel and White's influential three tiers cover only the knowledge gained from first-hand experience of an environment. But not even novices will enter an unknown environment with blank minds. They will be equipped with some strategic proce-dural knowledge gained from experiences of other environments. Such knowledge concerns default expectations about the structure of the environment, and enables already the application of rules or strategies for orientation and wayfinding. For example, first-time tourists to Paris may search for Notre Dame somewhere in the center of the city, just because all the cathedrals they know of are in city centers.

These orientation and wayfinding strategies are heuristics. Generally, heuristics are experience-based techniques for problem solving designed to avoid exhaustive search. In orientation and wayfinding exhaustive search can be extremely costly, and accordingly a number of heuristics have been learned and ingrained in wayfinding behavior. However, since these heuristics are based on experiences from other environments it is not guaranteed that they lead to an optimal solution. Jumping on a bus in front of the arrival hall at Charles de Gaulle airport that promises to go to a central city station might be a valid heuristic, but it is less optimal than taking the local train RER B to Saint-Rémy lès Chevreuse, which stops at Notre Dame after about 38 min.

One of these ingrained strategies is a piloting strategy in the urban maze, *least angle first*. Knowing the direction to a target landmark, for example a sighted church tower, a person will dive into the maze and choose at each intersection the street segment deviating by the least angle from the (now estimated) direction to the target [85, 86]. As a piloting strategy this heuristics is heavily relying on a landmark. However, also survey knowledge can be used in heuristics. For example, *longest leg first* on network-like survey knowledge may be interpreted as a heuristic strategy to identify a shortest route or a route of fewest turns. Finding a long leg has not only strong correlations with shortest connections or few turns, it also may be cognitively cheaper to process than the other two, or is more robust (for the complexity of computing the shortest, quickest or simplest routes see, e.g., [42, 45, 133]).

In summary, we can identify three forms of spatial knowledge contributing to a mental spatial representation (see also [14]):

1. Knowledge gained from exploring an environment (embodied experience).
2. Knowledge gained from exploring secondary sources such as maps, photos, or written or spoken words (embodied is only the experience with the medium, while the experience of the environment is indirect via a reading process).
3. Knowledge gained from experience with other, similar environments.

3.2 The Role of Perception

The human senses mediate between the physical environment and the representation of this environment in the mind. Wikipedia, for instance, states:[2] "A mental representation is the mental imagery of things that are not currently seen or sensed by the sense organs". But a prior perception by sense organs has led to the mental representation in the first instance, and accordingly, current perceptions interact with the mental representation. Spatial mental representations, we would assume by now, are based on the experience of distance and direction from locomotion, and on experiences of objects or events that are location-specific.

3.2.1 Landmarks Grab Attention

Kahneman [94], simplifying, distinguishes two ways of thinking, which he calls for brevity System 1 and System 2. System 1 is the one below conscious thinking. It is subconscious, emotional and automatic, and hence, fast compared to the other way of thinking. As many human skills become internalized, below conscious thinking, they fall into System 1. For example, locomotion (once the baby has learned to walk), path integration and mental rotation belong to System 1. Even acquisition of information triggered by external stimuli can happen subconsciously [121], in interaction with System 1. System 2 does conscious reasoning, and hence, is the slow system. For example, dialog about directions involves System 2. People are able to reflect how they argue while they argue about the route they choose in a given situation.

O'Regan has developed theories how people can form an image of the exterior environment of the body from perception (sensors) and feelings (responses to sensor readings) [161]. The question whether we can trust our perceptions is of course an old one. Already René Descartes, out of his methodological skepticism, formulated the famous quote *cogito ergo sum* (originally *je pense, donc je suis*, [41]), putting

---

[2]http://en.wikipedia.org/wiki/Representation_(psychology), last visited 3/1/2014.

the human ability to reason above his doubts about frequently deceptive perceptual sensations. The perceptions' deceptiveness still concerns research. Optical illusions demonstrate that our visual system already interprets retinal images before the conscious mind gets access. Purves and Lotto suggest that the visual system interprets stimuli based on experience, or on what has been learned to hypothesize as it had been true in many cases before [171]. Such a pre-processing is a prime example for System 1 thinking. One hypothesis applied by the visual system is the *grey world assumption*, the assumption that the average reflectance of an environment is grey. While this assumption in many cases provides useful interpretations, it can be misled by influences of illumination and reflection. Thus, Descartes' scepticism is appropriate, and O'Regan's quest is not trivial.

O'Regan claims that "an organism consciously feels something when it has at least a rudimentary self which has access consciousness to the fact that it is engaged in a particular type of sensorimotor interaction with the world, namely one that possesses richness, bodiliness, insubordinateness, and grabbiness" (p. 120). These four principles are:

Richness   The reality is richer than memories and imaginings.

Bodiliness   Vision and other senses have bodiliness, as any movement of the body produces an immediate change in the visual (or other sensory) input.

Insubordinateness   The sensory input is not totally controlled, or can change even if the body does not move because the external reality changes.

Grabbiness   Senses like the human visual system are alert to any changes in the sensory input.

The experience of landmarks is an involuntary act, one that does not involve conscious decision or choice. In this sense, landmarks must have properties that correspond to the grabbiness of human senses. They can have other, less grabbing attention properties as well, because of the richness of reality. In addition, a human will experience that this particular sensory input is related to a particular location in the environment instead of a particular body movement. It can only be repeated by reaching the same location. This aspect relates to O'Regan's insubordinateness.

Accordingly, in Kahneman's categories, learning an environment by landmarks involves System 1. The mind develops a representation of the environment that is independent from intellectual rigour or effort, and merely based on embodied experience (and thus, of course, depending on attention). The mental representation, however, is accessible to both, System 1 and System 2. People find their way in a known environment and orient themselves in the environment without an explicit involvement of System 2. During wayfinding, the interplay between expected sensory input from certain locations and actual sensory input confirming these expectations provides a feeling of being oriented. System 2, however, can access the mental spatial representation to explain, for example, the current orientation with respect to a few learned landmarks. Similarly, although the route of daily commute is travelled without conscious thinking, it takes System 2 to explain this route to another person.

Grabbiness is related to *affordance* [63]. If objects or events have to grab attention to become landmarks they must stand out in their neighborhood, or afford support in orientation, wayfinding and other spatial tasks. Since people continuously and mostly subconsciously maintain their orientation in an environment (another System 1 function), some external, perceivable properties must exude affordance to be used in short-term orientation. Other perceivable properties must exude affordance as long-term spatial anchor points. Again other perceivable properties must exude affordance supporting route planning, for example street signs or departure boards in the environment of a wayfinder. Affordance is task-bound or consciously purposeful. A wayfinder will study the departure board, but the cleaner will ignore it. And since the departure board grabs attention from travelers, and is at a central place for travelers, it may become a spatial anchor point, a landmark in their mental spatial representation helping with local orientation or route descriptions. For example, for tourists arranging to meet at 5 p.m. under the departure board of the city's train station makes perfect sense.

In his groundbreaking work on conflict Schelling brought up an example of non-verbal communication that relates to affordance:

> You are to meet somebody in New York City. You have not been instructed where to meet; you have no prior understanding with the person where to meet; and you cannot communicate with each other. You are simply told that you will have to guess where to meet and that he is being told that you will have to guess where to meet and that he is being told the same thing and that you will just have to try to make your guesses coincide. This problem showed an absolute majority managing to get together at the Grand Central Station (information booth), and virtually all of them succeeded in meeting at 12 noon ([188], p. 56).

3.2.2 Properties That Grab Attention

But what is it about some objects and events that grab this attention? While in principle the world is a continuous phenomenon, the human mind classifies visual and other perceptions of its environment and identifies discrete objects in this continuous environment. Visual perception alone has already self-organizing and holistic processing built in. This was discovered and described by Gestalt theory long before a neuroscientific understanding could be developed ([48, 200, 236, 237], but also [138]). Ehrenfels characterized the fundamental problem of Gestalt psychology with these words:

> Here we confront an important problem [...] of what precisely the given presentational formations (spatial shapes and melodies) in themselves are. Is a melody (i) a mere sum of elements, or (ii) something novel in relation to this sum, something that certainly goes hand in hand with, but is distinguishable from, the sum of elements? ([48], p. 250, translation from German following [200]).

Gestalt theory studied the basic laws of visual or aural perception. However, the laws can also be observed in thought processes, memories, and the understanding of time. It started from the observation that perception easily separates figures from ground by some fundamental rules of configurational qualities. These Gestalt rules are:

Proximity   Perceived phenomena near to each other are more likely perceived as Gestalt.

Similarity   Perceived phenomena similar to each other are more likely perceived as Gestalt.

Simplicity   Simpler explanations for a configuration are more likely to be selected by perception.

Continuity   Configurations in a linear order are more likely perceived as belonging together.

Closure   Closed configurations are more likely to be perceived as Gestalt.

Joint destiny   Perceived phenomena that move in the same direction are more likely perceived to belong together.

Phenomena—objects or events—experienced as landmarks must have strong (visual, aural, or other, but for humans predominantly visual) figure qualities. Considering these Gestalt rules, they must have strong local contrast by being dissimilar or far from the rest. To find out more about salient phenomena in mental spatial representations, Appleyard asked people in a survey for the buildings in their home town they could memorize best [8]. The collected buildings were characterized by Appleyard with respect to a set of properties. The correlation of properties with the number of nominations was used to identify the significant properties. He found:

Form properties   Inflow of people, contour, size, form, and visual attributes of the facade.

Visibility properties   Frequency of visibility, prominence of the view point, and nearness of the building to its view points.

Semantic properties   Intensity of use of the building (traffic), or uniqueness of use.

Appleyard observed that the more a building stands out the more likely it is that it comes to mind in such a survey. He found also that the correlation between the number of nominations and the significant properties shows both in local comparisons (i.e., for a neighborhood) as well as in global comparisons (i.e., for the city), which relates to our prior distinction of local and global landmarks.

More recent investigations to characterize the properties of landmarks are made by Sorrows and Hirtle [201] and Burnett et al. [19]. They also identified properties such as uniqueness, distinguished location, visibility, and semantic salience. Another obvious property should be permanence (e.g., [20], p. 67). Studies with children have shown that they also choose non-permanent objects, while adults do not [6, 29]. Sorrows and Hirtle [201] distinguished in particular:

Fig. 3.4 China merchants bank tower, Shenzhen, China: "The distinctive silhouette serves as a beacon for the residents of Honey Lake, Che Gong Miao, The East Sea gardens, and Shenzhen's west side" (CC-BY-SA 3.0 by *Daniel J. Prostak*, caption from Wikimedia)

Visual landmarks   Landmarks by their visual peculiarities
Semantic landmarks   Landmarks by their distinguishing use or meaning
Structural landmarks   Landmarks by their location in the structure of the
   environment

These categories are not exclusive. For example, a landmark can be eye-catching *and* culturally important. An object or event that is outstanding in more than one category should have a stronger total salience, in line with Appleyard's observations. One may call this amount of salience *landmarkness*. But if salience is a quantity, even a quantification of the landmarkness in each category becomes imaginable now.

Furthermore, the three categories should be applicable to all objects or events in human environments, natural or urban. Prototypical examples are buildings. Buildings can stand out by visual properties, by their use, or by their location. For example, the China Merchants Bank Tower in Shenzhen (Fig. 3.4) is standing out by its height, towering over other buildings in the environment, but also by its characteristic shape. People in Shenzhen use it as a widely visible landmark.

But have a look at another example (Fig. 3.5). This corner store is most likely not a global landmark in its environment as Guangzhou has multiple stores of this chain.

Fig. 3.6 Tuchlauben 19,
Wien: a building from the
fourteenth century, but so are
the left and right neighbor.
Yet this one is special

Nevertheless it stands out in its immediate neighborhood by (commonly known) visual characteristics (franchisees will ensure that stores are not too close to each other, to avoid cannibalism), by its use, and by being located at a corner. Hence it can serve as a reference point in mental spatial representations, and thus in route descriptions: "At the corner store turn left".

Despite the dominance of visual perception in human cognition, there are also landmarks of weak or no visual distinction. In Fig. 3.6, for example, it is hard to isolate the building in the center from its neighbors visually. The buildings form a visual unity. Even the buildings in the larger neighborhood are of similar age and style. However, this building shelters in one floor the oldest secular frescoes in Vienna. The rest of the building is still used as an ordinary apartment house, but the frescoes make it distinct in the eyes of the citizens of Vienna and many visitors.

Other buildings—despite potential visual and semantic salience—stand out just by their structural properties. For example, airports are typically known as nodes in the air travel network. People experience them mostly from the inside, where they more often than not look interchangeable. For example, Singapore Changi Airport is a node in this network experienced by millions of travelers per month. A travel agent telling her customer "At Changi, change to the flight to Frankfurt" may help with orientation and wayfinding. But usually it does not evoke or relate to the visual image of the terminal buildings. Train stations have a similar prominence in spite of low visual imaginability.

Similarly, other categories of objects or events can have this landmarkness. For example, in natural environments a prominent representative for landmarks is landform. Prototype landform objects are mountains, saddles or rivers, which all can be characterized by their visual, semantic or structural qualities. Mathematically they form singularities in the terrain height. Even slope alone is already observed as a peculiarity in learning, memorizing and communicating environments and routes. In orienteering a route description can be: "Walk uphill". The salience of slope is visual as well as structural (in terms of resistance, or increased physical effort to overcome this part of the route). In Lynch's categorization slope would form a gradual, but conquerable barrier. However, little literature exists on natural landmarks. For example, Brosset et al. report that landform is the second largest category of landmarks in route directions in natural environments [15]. In urban environments landform has not been systematically investigated. This may be caused by the availability of local, fine-grained objects to choose from (landform is of comparably coarse grain), or also by the fact that salient patterns of landform (e.g., steep slopes) are relatively rare in urban environments.

For urban environments Ishikawa and Nakamura have tested which categories of objects people use as landmarks [88], however in a narrow context. In their experiment people walked unfamiliar routes in an urban environment, and were asked to nominate the "objects that they thought were helpful as clues for navigation", or more precisely, "to imagine a situation where they use those landmarks to explain the routes to someone who visited the place for the first time" (p. 8). Participants were also asked to give reasons for their selections. Despite this narrow context, Ishikawa and Nakamura could show that people picked various objects such as buildings, signage, street furniture or trees. Also, since this was a reporting task to an experimenter, not an instruction task to a wayfinder, participants with a better sense of direction tended to select fewer landmarks. They also studied the physical properties of selected buildings, finding that facade area, color saturation, and age affected the selection, but that these parameters can differ between persons and routes. Denis et al. [40] also looked at object categories in verbal route descriptions of residents of Venice. They found that streets, bridges and squares were mentioned far more frequently than buildings. Lynch's collected sketches of some North American cities [128] would allow a similar quantitative analysis, which has not been done yet.

However, in each of these studies the environment itself (such as Venice with its bridges) as well as the assumed activity or purpose provide a specific context. Both context factors, environment and activity or purpose, influence the choice of object (or event) categories. Hence, the findings cannot be compared or averaged. For example, the literature is typically focusing on urban environments and wayfinding by walkers, which explains the prominence of object categories such as buildings or streets. In such a context, objects (or events) at this level of spatial granularity allow to anchor the decisions of a person with low ambiguity: "At the corner store turn left" works well, while "on [the hill] Montmarte turn left" does not.

Hence, the guideline for object or event categories being perceived or used as landmarks is actually the current context, or focus of the observer. The attention and intention of the observer regulates the affordance of objects.

3.2.3 What We Have Learned About Landmarks from Perception

Landmarks are the attention-grabbing objects in the environment. The experience, either first-hand or mediated, of these objects is stored in mental spatial representations anchoring the location because of its memorizable sensation. Since attention is context-dependent, also what is attention-grabbing is context-dependent.

3.3 Mental Spatial Representations

Generally, perceptions of any sort can lead to memorizable events ("the desert today was superb"). Memory, or a *mental representation*, stores primarily the properties of the experience. Landmark experience, however, is different. Objects or events are perceived as landmarks in relationship to their location. Thus, a *mental spatial representation* stores the experience together with its location. Furthermore, since the object or event was perceived in the context of moving in the environment, location is in the first instance related to the pose and heading of the ego, in a second instance related to what has been experienced before (the own trajectory), and thirdly to the larger knowledge acquired so far about the environment. This means, location is stored primarily in relation to other (known) objects or events.

This section approaches landmarks in mental spatial representations from two angles, brain and mind. While brain is frequently associated with matter, and mind with cognitive abilities, consciousness and personality, the distinction is not clear cut. Neuroscience explains gray matter by function as well, and, as mentioned before already, cognitive abilities are not necessarily all conscious. However, the observational approaches of the sciences are orthogonal. Neuroscience studies the brain from the perspective of cells and synapses, enzymes, and receptors at micro-level, and roles and activities in brain regions at macro-level. It applies invasive and non-invasive methods, both stationary due to limitations by technical equipment. Cognitive science, in contrast, studies cognitive capacities through observing people (or animals) behaving in situ or in response to external stimulation.

3.3.1 The Brain

The hippocampus is the region in the brain that has a central role in encoding and retrieving information for behavior guided by memory, not only for spatial behavior but for a range of different behaviors. However, just as the hippocampus is not solely occupied with spatial memory and abilities it is also not the only region in the brain involved in spatial memory and abilities. This makes mental spatial representations and reasoning one of the more complex areas for neuroscience to understand. Adding to the challenge, neuroscience can only rely on indirect (non-invasive) observations on humans. Recent technology such as functional magnetic resonance imaging (fMRI) is observing oxygen levels in blood, which may be correlated with cell activity but must not be identified with it. Patients with brain lesions provide further, but also indirect insight into the working of the brain [36, 181]. Whether cell activity of rodents, observed by invasive procedures, can be taken to explain human capacities is open for debate. Furthermore, several types of brain cells seem to be involved in mental spatial representations. With one of them, *grid cells*, being discovered only less than a decade ago [59], it may be no surprise that there is still uncertainty and speculation about the mechanics of mental spatial representations [46, 174].

Very early, and based on behavioral experiments with rats, Tolman established facts for a direct spatial representation in animals' long-term memory:

> We assert that the central office itself is far more like a map control room than it is like an old-fashioned telephone exchange. The stimuli, which are allowed in, are not connected by just simple one-to-one switches to the outgoing responses. Rather, the incoming impulses are usually worked over and elaborated in the central control room into a tentative, cognitive-like map of the environment ([216], p. 192).

Some 30 years later O'Keefe discovered *place cells* in the hippocampus of rodents. These cells fire when a rat returns to a place it had visited before [158]. The book he wrote with Nadel [159] still suggested the existence of a *cognitive map*. It was too tempting to believe that the brain stores the experiences at particular locations in a map-like fashion. By now, however, it has become clear that place cells are not the only cells involved in mental spatial representations (e.g., [173]), and that spatial abilities are formed by complex interactions between various regions in the brain.

Place cells, or fields of place cells, show firing patterns depending on location. But how does a rodent's brain know? In principle there are two ways for an animal to locate itself: One by path integration, i.e., sensorimotor stimulations, and the other by perception of external cues. Since place cells fire even when an animal moves in the dark, current thinking is that hippocampal representations are primarily driven by path integration. Since path integration is accumulating uncertainty over time, external cues may help to stabilize the localization, but are clearly second order.

In contrast, *head direction cells*, monitoring the direction of the face, have been shown to be sensitive to visual external cues [209, 210]. This is no surprise since stationary visual external cues provide a stable reference frame for turning the head. However, head direction cells also operate in the absence of light, supported by the vestibular system, but then, similar to place cells, with drift over time. Head direction cells are found in many regions of the brain, including the postsubicular cortex.

Grid cells [59] seem to code location in a regular tessellation, similar to an internalized coordinate system. Since the tessellation covers space, firing patterns are sufficient to decode location in space [74]. Grid cells are found in the medial entorhinal cortex, an informational input into the hippocampus. There must be a mapping of locations identified by firing of the grid cells and locations identified primarily from path integration in the place cells, and further associations between place and events to form memories (ibid.).

Another pathway into the hippocampus deals with object recognition establishing some contextual information about location. With contextual, especially visual information provided, the brain is capable of allocentric spatial reasoning, probably in the posterior parahippocampal cortex. Ekstrom et al. [49], for example, found cells in the hippocampus, parahippocampal cortex and other areas that were responding to location, but in addition also cells that fired dependent on the visual external cues the person viewed:

> We present evidence for a neural code of human spatial navigation based on cells that respond at specific spatial locations and cells that respond to views of landmarks. The former are present primarily in the hippocampus, and the latter in the parahippocampal region. Cells throughout the frontal and temporal lobes responded to the subjects' navigational goals and to conjunctions of place, goal and view (p. 184).

If grid cells form an internalized coordinate system their resolution and number becomes interesting. The resolution determines the smallest variation in location that can be distinguished in firing patterns, and their number determines the size of the environment that can be represented. Just as a technical equivalent: If the surface of the Earth should be represented in a tessellation of square elements of 1 m edge length it needs $40,000,000^2$ or $1.6 * 10^{15}$ elements. It is estimated that the human brain overall has about 100 billion neurons, thus, human spatial memory must be differently organized to remember things such as where the keys have been left or how to travel to Sydney. It needs hierarchic representations. In fact grid cells show properties that establish multi-resolution memory. First, different areas of similar-sized grid cells represent the same environment, but with a random offset of their grids. By nesting, even a small number of neurons can represent a fine level of granularity [139]. Secondly, the scale of grid cells varies along the entorhinal cortex. For rats, grid cells have been found representing distances of about 25 cm in their dorsal-most sites to about 3 m at the ventral-most sites [16]. And finally, it has been shown that when certain channels are knocked off the brain produces a coarser scale spatial memory [65].

In addition to a representation of location the brain shows two additional types of spatial memory. One is episodic. Sequences of place cells form a route memory that can be imagined and mentally travelled at ground perspective. Burgess et al. [18] put it more precisely:

While processing of spatial scenes involves the parahippocampus, the right hippocampus appears particularly involved in memory for locations within an environment, with the left hippocampus more involved in context-dependent episodic or autobiographical memory (p. 625).

The other one is for survey-like, from-above visual imagination and manipulation of environments. Shelton and Gabrieli [196] have observed people viewing an environment in each of the two perspectives with fMRI. When comparing the brain activation during route and survey encoding they found that both types of information recruited a common network of brain areas, but while survey encoding recruited a subset of areas recruited by route encoding, route encoding, in contrast, recruited regions that were not activated by survey encoding.

Similarly, routine behavior in well learned environments may be stored as schema knowledge outside the hippocampus [219]. This means, people in spatial decision making situations may follow schemas (mental shortcuts) rather than analysing their mental spatial representations.

In all likelihood a person will not experience every square-meter of the Earth in their life-time. But there will be a difference between the traveling range of the peasant in the middle ages, who rarely left the district, and a global nomad of the twenty-first century. The global nomad at least will appreciate schema knowledge after experiencing that all airports look alike. Relph called them even 'no-places' [175]. On the other hand, Maguire et al. found that spatial memory shows plasticity in response to environmental demands [130]. They compared the brains of London taxi drivers with control participants who did not drive taxis. It turned out that the posterior hippocampi of taxi drivers were significantly larger, and this observation correlated with the amount of their professional experience.

The involvement of the parahippocampal region in both spatial memory as well as object or scene recognition is particularly interesting in the context of landmarks as used for visual navigation. Janzen and Turennout [92] have investigated combinations of skills:

Human adults viewed a route through a virtual museum with objects placed at intersections (decision points) or at simple turns (non-decision points). Event-related functional magnetic resonance imaging (fMRI) data were acquired during subsequent recognition of the objects in isolation. Neural activity in the parahippocampal gyrus reflected the navigational relevance of an object's location in the museum. Parahippocampal responses were selectively increased for objects that occurred at decision points, independent of attentional demands. This increase occurred for forgotten as well as remembered objects, showing implicit retrieval of navigational information (p. 673).

We will come back to this encoding of navigational relevance; in Sect. 3.2 we have called it structural salience. In addition, Janzen and Turennout demonstrated later that good navigators show even a consolidation effect in their spatial memory. Their activity in the hippocampus increases when recognizing objects along routes learned a while a ago, compared to routes traveled just now [91]. Furthermore, Maguire and colleagues have investigated whether a specific human navigation system exists [129, 203]. Their experiment was based on fMRI while tracking participants in virtual environments. They could identify three specific brain regions supporting navigation, which together seem to code the proximity and direction to the goal. The three regions were the medial prefrontal cortex and the right entorhinal region, neuronal activity in one of them positively correlated and in the other negatively correlated with goal proximity, and the bilateral posterior parietal cortex, where activity was correlated with the egocentric direction to goals.

With such a differentiated picture it makes sense to assume that different navigation tasks access different spatial abilities. Hartley et al. observed brain activity in people finding their way in an unfamiliar (virtual) environment and in people following a familiar route in another (virtual) environment [75]. The wayfinders showed higher activities in the anterior hippocampus, while the route followers showed higher activities in the head of caudate. The prior coincides with the prior assumption that the hippocampus is involved in place learning, and provides response to location within a spatial representation. The latter is consistent with an assumption of the caudate supporting action-based representations, and providing fast response to actions instead of locations. Their observations suggest that (at least) two representations are available for navigation, an action-based which is more efficient in learned environments and a location-based which is more efficient in unknown environments.

A review of behavioral and neuroscientific findings in rodents and humans by Chan et al. brought up that environmental objects can act as landmarks for navigation in different ways [23]. They proposed a taxonomy for conceptualizing object location information during navigation. This taxonomy consists of:

- Objects as visual beacons for navigation indicating a nearby en-route target location.
- Objects used as associative cues indicating a nearby location with an associated navigational action.
- Objects as visual cues to maintain or regain orientation.
- Objects used as an environmental reference frame for navigation, which are geometric properties of larger objects that can provide a frame for organizing spatial information, such as alignment operations (for example, rats seem to prefer geometric cues over object cues for orientation [25]).

The distinction between smaller objects—visual beacons or associative cues—and larger objects, or rather object geometries such as walls, becomes even more relevant with research testing whether one of them is preferred. Hartley et al. [76], for example, have geometrically altered the boundaries of a (virtual) environment between two visits of participants. The first time participants encountered a cue object in a simple rectangular enclosure and a distant visual cue for orientation. The second time, after a brief break outside of the environment, participants were brought back and asked to mark the place where the cue had been. On some trials the geometry (size, aspect ratio) of the arena was varied between presentation and testing. Hartley et al. report:

> Responses tended to lie somewhere between a location that maintained fixed distances from nearby walls and a location that maintained fixed ratios of the distances between opposing walls. The former were more common after expansions and for cued locations nearer to

the edge while the latter were more common after contractions and for locations nearer to the center. The spatial distributions of responses predicted by various simple geometric models were compared to the data. The best fitting model was one derived from the response properties of place cells in the rat hippocampus, which matches the proximities $1/(d + c)$ of the cue to the four walls of the arena, where d is the distance to a wall and c is a global constant.

Also, the geometry of the arena seemed to have served as a weak cue for orientation. Overall people seemed to combine two strategies: matching the distant visual cue for orientation, and representing proximity to the walls of the arena consistent with path integration.

Knauff added to this discussion [104] by arguing for a stronger separation of mental visual imagery and spatial reasoning. He refers to the long tradition in cognitive science of thinking of a pathway for visual object identification—the processing of visual properties such as shape, texture and color—and a pathway for recognizing where objects are in space. He also cites Landau and Jackendoff [114], who observed evidence in language encoding for a non-linguistic, cognitive disparity in the representation of *what* and *where*. Similarly, one could cite the visual sense with its separate predispositions for the detection of location and the recognition of objects [13, 138, 191]. Here Knauff suggests that "human reasoning is based on spatial representations that are more abstract than visual images and more concrete than propositional representations" (p. 16), claiming that spatial representations are not restricted to a certain format, and integrate different types of information while avoiding excessive visual detail.

Now, taking Knauff's suggestion seriously, a particular role for landmarks opens up. Since landmarks have both properties, the visual imagery of object identity and the anchoring to location, landmarks do form the bridge between mental visual imagery and mental spatial representations. This location-object binding will be useful in both, the subconscious, automatic cognitive processes (System 1), such as self-localization, and conscious cognitive processes (System 2), such as searching for a landmark provided in a verbal route description.

3.3.2 The Mind

Now let us focus on the cognitive capacity to form and maintain representations of the spatial environment, and to recall from these representations for reasoning or sharing knowledge with others. We will see that landmarks play a central role in all these processes. We will concentrate on cognitive studies; Sect. 3.4 will be dedicated to sharing knowledge with others.

3.3.2.1   What Developmental Psychology Has to Say on Mental Spatial Representations

Developmental psychology's interest in spatial cognition was kicked off by pioneers creating two schools of thought in parallel. One school is based on the work of Piaget and Inhelder [166], who assumed an innate desire and ability of children to learn from perceptions. The other school is based on the sociocultural approach of Vygotsky [229], who assumed a greater role of education for the cognitive development of children. In the tradition of Piaget and Inhelder, *cognitive psychology* studies mental representations of space and revisions of these representations, while *neuroscience* studies the brain activities in perceptual integration and spatial problem solving. In the tradition of Vygotsky, *cognitive anthropology* and *linguistics* study cultural differences in spatial conceptualizations and spatial communication. These conflicting approaches are still noticeable, especially since the representative communities continue in disagreement, as Newcombe et al. mention:

> Do infants develop into competent adults in a protracted course of development propelled by interactions with the physical environment (as Piaget thought)? Or do they develop due to social interactions, linguistic input, and apprenticeship in the use of cultural tools such as maps or the use of star systems (as Vygotsky thought)? Or are they actually equipped from the beginning with core knowledge of objects and space, later augmented by the acquisition of human language (as argued in the past few decades by Spelke)? ([156], p. 565).

At another place, Newcombe and Huttenlocher offer a perspective for reconciliation:

> Spatial development is an excellent candidate for a domain with strong innate underpinnings, due to its adaptive significance for all mobile organisms. Piaget, however, offered a developmental theory in which innate underpinnings of spatial understanding were rather humble, hypothesizing that simple sensorimotor experiences such as reaching are the departure point for a gradually increasing spatial competence that emerges from the interaction of children with the world. His approach dominated developmental thinking about space for several decades, and inspired a great deal of research, some lines of which are still active. Recently, however, there has been a return of interest in a more strongly nativist theory in which infants come equipped with specific knowledge of the spatial world, along with similar specific understandings of domains such as language, physical causality, and number ([155], p. 734)

Gopnik suggests a developmental process based on theory theory (in other contexts theory theory is also called folk theories). An infant, reaching out, collects embodied experiences and tests with these experiences its first, primitive theories about space. Causality, or at least probability, learned from repetition will lead to revisions and the development of increasingly mature theories [70, 71, 165]. Theory building mechanisms can be *specialization* by adding a constraint to a theory limiting it to special sorts or cases, *generalization* by removing such a constraint if a theory has been found to be a special case of a more general theory, and *dynamic weighting* between theories to assign importance to favored theories [224]. Based on these mechanisms the infant will learn, for example, that some things are out of reach, or that some things can be stacked, or that all the perceptual stimuli caused by a car form a whole, and will move together when the car moves. Thus, theory theory is about concept formation [132], and generally related to space (anything external to the infant's or young child's body). A significant portion is about vista, environmental and (later) geographic scale [146], and thus about localization, orientation and navigation.

Piaget and Inhelder [166] observed that children at certain ages approach spatial problems systematically different from adults. They postulate stages in the development of spatial abilities. The first stage, from birth to age two, is a sensorimotor stage where children form first notions of space based on their visual, aural, gustatory, olfactory or tactile perception, and their sense of movement (proprioception and equilibrioception) in a strictly egocentric frame of reference.

Later, at the age of two to seven (the pre-operational stage), children note the topological properties of objects in the environment, and consider properties such as separation, continuity and connectedness, closure, containment, and proximity (the latter is, strictly speaking, not a topological property). Children at this age can make qualitative judgements. A child can recognize which one of two lines is longer at an age long before it can estimate or measure a quantitative value of length [166]. Also at this stage transitions from egocentric to allocentric frames of reference can be observed. This transition is facilitated by landmarks. When Acredolo [1] led children of this age group into a room and asked them to search for a trinket they learned the room in an egocentric manner, but once landmarks were present in a room, these children did no longer rely on an egocentric search heuristic but considered the location of the landmarks as well. But much earlier than that, demonstrated at the age of 6 month already, infants can associate markers (landmarks) with an expected location of an event [179].

Only after this phase, at the age of seven to eleven (the concrete operational stage), children develop an understanding of projective properties of space, of distances and angles, and start thinking logically. From the age of eleven onwards (the formal operational stage) children develop abstract thinking, learn Euclidean properties of space, and order along dimensions of space.

Piaget and Inhelder's observations underscore that the notion of topology is sufficient for forming spatial schemata and solving many spatial problems including navigation. Metric notions, in this regard, are mostly needed for refinement, for example, for navigating the shortest route. Accordingly, mental spatial representations develop notions of topology long before, or treat topology as more fundamental than metric notions. We will see later a reflection of this order in communication about space.

Research has also investigated how children develop skills in reading maps, i.e., external spatial representations. Liben and Myers provide a review [122]. Children have to learn that maps are representations: "the child must first have the basic understanding that one thing can be used to stand for something else" (p. 198), and "a basic challenge [...] is for children to learn to differentiate which qualities of the symbols carry meaning and which are simply incidental objects of the symbols" (p. 199). The child's interpretation of the map is linked to their developmental stage or age. Correspondences between the environment and the map have to be established, and some of them are of topological nature, some of them metric. The metric ones are understood only by older children. Age differences show also in the ability to extract relevant or appropriate information and to mentally rotate a map to establish correspondences. Furthermore, beyond vista space children must establish correspondences with their memories, or their knowledge of environmental or geographic space.

Blades [11] focuses on children's strategies to extract appropriate information from maps, and to locate themselves on the map. He finds that children from the age of 3 years have some appreciation of correspondences, and from 4 years of age they start using a map and can locate themselves—which is a qualitative or at most comparative procedure after establishing the correspondences, and does not require metric notions. But only older children can use maps for more elaborate strategies such as finding places or following routes because these tasks go beyond vista space and require logical and abstract thinking.

One particular challenge of establishing these correspondences raises from the perspective taking[3] of the map. The map is built for a purpose, with other words it is a focused and selective representation. The purpose dictates scale, selection of map objects (classes and individuals), and cartographic representation including symbology and coloring. The same environment can be represented by entirely different maps. Especially landmarks, the anchor points for spanning a frame of reference in the map reading process, can appear in a large variety of representations. The point-like landmark—the landmark in Lynch's sense—may be represented by a symbol, by a simplified, potentially perspective view, by an extended region, or by text. Since other elements can have landmarkness as well, such as rivers, streets and intersections, the process of establishing a frame of reference amounts to a complex procedure already in the mind of a child.

3.3.2.2 Frames of Reference

Mental spatial representations, just as external ones, are based on spatial frames of reference. The role of a frame of reference is facilitating an unambiguous way to locate things in a space. A frame of reference is established by its datum. Mathematically the datum comprises an origin location and direction (short: *origin*) from which distance and direction measurements are made.

For a cognizing individual the datum can be an oriented object in the real world, defining location and direction. Two kinds of objects come to mind: The own body (an *egocentric perspective*), or another oriented object (an *allocentric perspective*) [17, 99, 106]. Taking my body as origin I can visualize that "the library is right of me, not far", which relates the library to my body, or describes its location with respect to my body's location and heading. Using an egocentric frame of reference has some challenges, despite being the frame of reference babies

---

[3]The term *perspective taking* is used here deliberately in the most general sense of focusing on a situation, selecting relevant objects and relations, and then choosing a geometric perspective. In contrast, linguistics understands only the latter as perspective taking [118].

develop first. One challenge lies in the ambiguity about the orientation of the body, where chest, head and viewing direction can be, and frequently are, not aligned. If the body is moving, then the heading direction can be non-aligned with any of the prior ones. Another challenge, and a more significant one, lies in the body's mobility, which requires constant updating of a representation of all the relationships of the body with the objects in the environment. This updating is actually much more costly than maintaining an allocentric, stable representation of the location of stationary objects and only updating the body's location and orientation in this representation. I can mentally visualize: "The car is parked in front of the church", which has no reference to my body at all, but instead refers to the inherent orientation of the church, which has a front side by design. When I move or turn, the relationship between car and church remains stable.

Updating my location in an allocentric frame of reference can happen in two ways. One was mentioned before, path integration, sometimes called *dead-reckoning*. This updating process through path integration is independent of the configuration of objects around. Path integration provides an update only with respect to an initial location. Loomis et al. investigated the ability of path integration by homing experiments with either blind or blind-folded humans [125]. Similarly to technical dead-reckoning systems, which are equipped with inertia sensors, path integration quickly accumulates error.

The other updating mechanism is *piloting*, orientation by recognizing landmarks in the world, and establishing the own body's relationship to these landmarks in their known configurations. Piloting, therefore, has no accumulation of errors. But it relies on the presence of recognizable landmarks in the proximity. Unavoidably piloting has to deal with gaps between experiencing these landmarks, and path integration can help bridging these gaps.

As early as 1913 Trowbridge postulated two frames of reference, one that he called egocentric but actually used cardinal directions instead of body orientation, and a *domicentric* that puts *home* (lat. *domus*) in the center [218]. He thought within range of home mental spatial representations will be domicentric, and what goes beyond will be computed by polar coordinates, a form of path integration in the frame of the orientation of the Earth, and hence an allocentric perspective. Trowbridge must be lauded for postulating a mental spatial representation so early in the twentieth century. His classification is no longer supported, although our thought experiment (Chap. 2) used 'home' as the first point of reference.

The thought experiment (Chap. 2) had used 'home' as an allocentric first point of reference, later replaced by a pole together with a reference direction (see also Fig. 3.7). But with more and more landmarks in the experiment's environment a single datum point can be traded for configurations of landmarks, at least for local

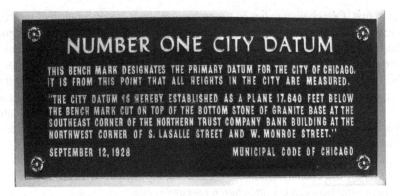

Fig. 3.7 A datum point in the real world (CC-BY-SA 3.0 by *Cosmo1976*, modified)

reference frames. In a perfectly allocentric perspective, "between the two lakes" inherits an orientation from the locations of the two (non-oriented) lakes. "Between dry cleaner and bakery", an example with the same structure but oriented objects, also inherits the orientation from the locations of dry cleaner and bakery, but is visualized probably more like "between two locations on the street, one in front of the dry cleaner, one in front of the bakery".

Configurations of landmarks as alternative manifestations of a frame of reference appear quite attractive to explain mental spatial representations as well as spatial communication. A configuration of stationary objects spans a vector space (object-to-object vectors), and everything else can be located within this framework in relation to elements of the defining configuration. The configuration needs to be updated only when changes to the objects are experienced: additions, deletions, or modifications. Additions and deletions in the world will require adding or deleting nodes in the representation and updating the object-to-object vector space to remain a consistent configuration, while modifications only concern the properties stored with a node in the configuration. Furthermore, these updates are local. They concern only the object and its relations (vectors) to neighbored objects.

Configurations of landmarks enable only to describe locations relative to the landmarks in the configuration, i.e., locally. For many tasks such as locomotion, object manipulation, speed estimation and spatial updating such local frames of reference are sufficient since they involve proximal objects. Even for tasks extending beyond proximal objects, such as wayfinding, a global frame of reference is not essential. A person living in a city can find their way around without knowing where North is, or where the coast is, just by knowing the vectors of relations between the landmarks in the configuration.

However, a globally oriented configuration can simplify wayfinding strategies. For example, an allocentric visualization "the library is North of here" refers to an external, global frame of reference by using a cardinal direction. The reference provides a wayfinder with global orientation and allows applying the least angle heuristics to North, which relieves from considering more detailed configurational knowledge in the planning process (actually, it does not require further configurational knowledge at all). In order to supplement mental spatial representations by anchoring configurational knowledge, global orientation must be based on phenomena that can be experienced by the human senses. Among these perceivable phenomena are the position of the sun or of the stars, of winds or currents, and potentially the Earth's magnetic field [126]. However, while a sense for the magnetic field has been proven for some animals [57, 96, 230], evidence for a human sense remains inconclusive [72].

But even semi-stable azimuthal references are considered by humans in spatial problem solving. Geographic objects that are either sufficiently far away or sufficiently large that the azimuth does not change significantly during a spatial task can serve as an azimuthal reference also, although with higher error margin. According to our prior classification these objects belong to the global landmarks. They can be the peak of a mountain visible on the horizon (sufficiently far away), the direction to a neighboring city as learned by traveling (sufficiently far away), or a coastline (sufficiently large).

Datum location and orientation are arbitrary choices, and between the different choices a linear transformation by a shift and rotation provides compatibility. This arbitrariness of choosing a datum is a fascinating property. For reasons of cognitive efficiency it is actually chosen according to a task at hand. The mind does not have to retrieve the whole configurational knowledge from long-term memory into working memory for a task at hand, but can be selective and reduce the cognitive load significantly to the landmarks and vectors that are relevant. We may call this ability to focus. As advantageous as focusing is, there is a price to pay. As Montello has put it: "There is always some aspect of location or heading that a person does not know precisely. In other words, everyone is potentially disoriented to some degree at all times" ([149], p. 264).

3.3.2.3 Hierarchies

Assumptions of a 'map-like' mental spatial representation were expressed since Tolman [216] and maintained for some time (e.g., [119, 159]). A more carefully phrased explanation was presented by Downs and Stea [44]:

> Cognitive mapping is a process composed of a series of psychological transformations by which an individual acquires, codes, stores, recalls, and decodes information about the relative locations and attributes of phenomena in his everyday spatial environment. A cognitive map is the product of this process at any point in time.

They did no longer make a direct claim that this product is a single integrated picture. Later Tversky made a point that the term *map* can only have metaphorical meaning for an actually more complex mental spatial representation, one that has also some properties of a collage structured by some spatial models of relations between objects [221]. For good reasons we therefore stick in this book with the neutral term *mental spatial representation*. There is no single, homogeneous pictorial 'mental map', or synonymously, 'cognitive map'.

Mental spatial representations have one property that can only be explained in a relational way. They are hierarchic. Hierarchic structures have notable advantages. One advantage is that a search for individual entities is significantly faster when applying hierarchial heuristics compared to exhaustive search. Readers familiar with spatial database structures see parallels to indices and hierarchical data structures [185]. The other advantage is that different spatial tasks are carried out more efficiently at different levels of a hierarchy. For example, wayfinding requires access to the street network of an environment, while coordination of locomotion requires a far more detailed representation of a significantly smaller environment [215]. Selective access to mental spatial representations is therefore essential for cognitive efficiency.

Typically, each coarser level of a hierarchy is derived by abstraction from the next finer level [84]. This is not the case for mental spatial representations. For example, geographic space, in contrast to environmental space, can only be learned from secondary sources such as maps. The mental spatial representation of, say, a country's boundaries, or its neighbors, is not derived from extensive locomotion. Distant (global) landmarks can be learned from the distance, without travelling or knowing more detail about them. Hence, mental spatial representations are not filled at every level. Even the notion of 'levels' is questionable, but useful to imagine configurations of landmarks in a vector space. At a coarse level of detail these landmarks can be the countries of Europe, with Austria to the east of Switzerland. At a fine level of detail these landmarks can be the furniture in my living room, with the table in front of the couch. Loosely, we expect similar things to be collected in a level of a hierarchy.

Levels are connected by links for reasoning purposes. These links establish relationships, and depending on the type of relationship different hierarchies can be constructed. Spatial hierarchies cater for both, the *selection* of objects that are retrieved from memory as well as the *level of detail* with which they are retrieved. Readers familiar with cartography may be reminded of zooming. Zooming through map series is supported by cartographic generalization techniques such as simplification, selection and aggregation. In fact, mental spatial representations maintain complementary hierarchies based on these three operations, a hierarchy by *detail* based on simplification, a hierarchy by *salience* based on selection, and a hierarchy by *granularity*. Let us have a closer look at these hierarchies, and their empirical evidence.

Hierarchies by Detail

Landmark hierarchies by detail are rather subtle. They relate to the nature of landmarks, which can stand both for a location as well as an object. In the prior function they provide an anchor in a mental spatial representation. Similarly, Lynch [128] characterized landmarks as external reference *points*. In the latter function they provide memory for the shape and visual image of an object, for tasks such as object recognition. Landmarks seem to be able to bridge between the most abstracted form of an object as a purely spatial reference, and more detail of a visual image. For reasons of cognitive efficiency the mind may only recall what is just needed. For example, wayfinding according to a plan "at the corner store turn left" requires first to find a store, which can be based on a fast low-detail schema, and only when a store has been identified in the environment a detailed visual image from memory is required for object disambiguation.

Hierarchies by Salience

We encountered already global landmarks and local landmarks, landmarks that stand out in a larger environment—such as the Eiffel Tower in Paris—and landmarks that stand out only locally—such as a corner store along a street, or ATMs in the entrance area of a mall. The coarse classification hints to asymmetric relationships. For example, "the ATM at Eiffel Tower" may be accepted, but "Eiffel Tower at the ATM" is not. One is more salient than the other, even dominant.

Thus, configurations of landmarks (a 'layer' in a salience hierarchy) comprises similar objects, but similarity is based here on salience, not size or level of detail. Selection and weighting, we have argued above, are linked to the embodied or mediated experience of the environment. We mentioned already the following factors contributing to salience: The figure-ground contrast, the relevance of the location, and the frequency of encounter or prominence. The more salient landmarks are the more easily recalled objects of an environment [8]. If some objects are more easily recalled into working memory they must also be earlier available for processing. Besides of priming the mental spatial representation, they also serve as available reference locations for other objects nearby ("at the Eiffel Tower").

The first evidence for such reference points in mental spatial representations was provided by Sadalla et al. [184]. They asked participants for distance judgements between pairs of reference- and non-reference points. In their words:

> Since a reference point is regarded as a place that defines the position of other adjacent places, it follows that other places should more easily be seen in spatial relation to a reference point than vice versa [...]: Adjacent places should be judged closer to reference points than are reference points to adjacent places (p. 517).

One of their tests, for example, demonstrated different judgements for questions like "Is the newsagent's close to the hospital?" and "Is the hospital close to the newsagent's?", with the hospital assumed to be a reference object. Also, distance judgements to reference points are made faster than to non-reference points, and direction judgements to other reference points are made faster when at a reference point. Adding to these observations, Allen [2] found that distance estimates across object clusters around reference points were judged to be consistently longer than distances of the same length within such clusters.

Couclelis et al. [31] concentrate then on the link of objects with their reference object, i.e., the link of a location characterized by a landmark and its *reference region*. They call it the *tectonic plate* hypothesis. According to the tectonic plate hypothesis landmarks anchor distinct regions in mental spatial representations, called the *reference region* of a landmark. Spatial cognition research does not yet offer a good understanding of the size of these reference regions, or tectonic plates [51]. If in a configuration of similarly salient landmarks each object stands for its neighborhood one assumption is that the neighborhoods form a partition of space. For example, a mental visualization of "at the church" could include every location that is closer to the church than to any of the other salient objects around [238]. However, if the configuration changes such a partition must change as well. Winter and Freksa [238] also demonstrate that context can be captured by contrast sets of landmarks, which means that the configuration becomes context-dependent, or reference regions change between contexts. Furthermore, salience of entities is also weighted, and the influence of different weights on reference regions is also not clear.

The distinction between reference points, or landmarks, and non-reference points implies at least a two-level salience hierarchy. At least in theory there is no reason to limit this thought to a two-level hierarchy only. While we are not aware of any research documenting the observation of zooming through multiple levels of salience, at least anecdotal evidence suggests their existence. For example, we can easily distinguish global landmarks—those shared by everybody, tourist and local alike, local landmarks—those short-term tourists would not know but which are shared by locals, landmarks of specific groups among locals—e.g., those shared within family, and private ones.

Hierarchies by Granularity

We are all familiar with containment hierarchies. It is the way we have learned geography. Continents are aggregations of countries, and countries of states or counties. Accordingly, western postal addresses locate an addressee on street level, then on city level, then on country level. Such a hierarchical structure is established by part-of relationships from an entity at one level to an entity at the next coarser level of granularity. Salt Lake City is part of Utah, and Utah contains Salt Lake City. Mental spatial representations reflect these hierarchies by containment relationships. Developmental psychology has already documented the fundamental role of topology for cognitive spatial abilities (Sect. 3.3.2.1). Other literature identified a number of classifications of spatial granularity that are motivated by human conceptions of space (for reviews see [56, 177]).

A hierarchy by granularity, based on aggregation and abstraction, is independent from hierarchies on detail or on salience. Two objects of the same level of granularity can have very different salience. For example, the Guggenheim Museum in New York is more salient than the buildings next to it along Fifth Avenue, which are nondescript apartment buildings. Even more, an object can be more salient than its container object, i.e., salience does not accumulate with aggregation and abstraction automatically. For example, more people will confirm to have experienced (physically or mediated) the Champs-Élysées than the eighth Arrondissement of Paris, of which Champs-Élysées is part of.

Fig. 3.8 What is further east:
San Diego or Reno?

The classical experiment to provide evidence for this type of hierarchy has been made by Stevens and Coupe [206]. They asked participants to estimate the direction between pairs of cities that are located in different states or countries. For example, they were asked to estimate the direction from San Diego (California) to Reno (Nevada). Since the state of Nevada is east of the state of California, most participants made biased direction estimates to the east. But the direction from San Diego (California) to Reno (Nevada) is actually north-northwest (Fig. 3.8). Running the experiment for multiple pairs of cities, the only explanation for the consistent distortions the participants made in their estimates is by assuming hierarchical reasoning.

This hierarchic reasoning strategy can be illustrated by many more examples. Ask yourself (or your friends) which city is further North, Vienna or Munich, keeping in mind that Austria is south of Germany. Which city is further east, Vienna or Prague? Many people still remember Europe's division into a western and eastern bloc, and Prague being located in the Eastern Bloc. In contrast, most people have long forgotten or never learned that Prague was once (twice) capital of the Holy Roman Empire (from 1346 to 1437, and from 1583 to 1611), and hence, is quite centrally located in Europe. And which city is further south, Adelaide or Melbourne, given that Adelaide is the capital of South Australia? If you are more familiar with other regions in the world you will find other pairs of cities.

Stevens and Coupe point also to the efficiency gains of hierarchical reasoning. If a relationship can be stored at a coarser level of granularity then all the relationships between pairs of objects contained in both aggregates do not need to be stored. The adopted strategy in the particular example seems to be:

- San Diego is in California, Reno is in Nevada (lifting the problem to a coarser level of granularity).
- Nevada is East of California (reasoning about directions at this coarser level of granularity).
- Ergo, Reno must be East of San Diego (projecting the reasoning result to the finer level of granularity).

Look at exhaustive search in contrast. An exhaustive search for the direction from San Diego to Reno would require to activate in working memory the vector space of cities in the western part of the USA, and then to determine by vector addition the direction between the two named cities. No doubt that this exhaustive search is possible for the mind. A person living at the west coast of the USA may have learned the configuration of cities from travelling around extensively or from studying maps. But the cognitive load is significant, compared to lifting the problem up to the coarser level of granularity.

But the experiment also demonstrates the price to pay. A heuristic is a mental shortcut. It has become part of System 1 because it provides satisfactory solutions in general, whilst accepting that it does not always guarantee the correct (or optimal) solution [64, 94, 220].

Inspired by the above experiment, Hirtle and Jonides [83] hypothesized that mental spatial representations of inhomogeneous distributions of objects are also hierarchic. Clusters exist in the real world. For example, buildings are dense in cities and scarce in rural areas, cities are dense in populated countries and absent on sea, and so on. People judge distances between these clusters differently than within clusters, and thus, they form tree-like mental hierarchical structures formed by these clusters, even in absence of boundaries or barriers as in Stevens and Coupe's experiment.

Moreover, McNamara [141] studied how people learn environments and then make judgements from memory. He presented his participants either a physical environment or the same environment represented on a map. Both presentations had a hierarchic structure. Objects in the physical environment were laid out in regions marked on the floor, and correspondingly the map space was partitioned into regions. After a learning phase participants had to engage in three tasks: item recognition, direction judgements, and Euclidean distance estimation. The results from all three tasks were sensitive to whether objects were in the same region or in different regions, which is a clear indication of a hierarchically structured mental spatial representation of a learned environment.

Thus, there is strong evidence for a hierarchic organization of mental spatial representations by granularity, more precisely in containment hierarchies.

3.3.3 What We Have Learned About Landmarks from Mental Spatial Representations

In summary, landmarks appear to have a number of functions in mental spatial representations. They form anchor points in these representations, spanning a spatial reference frame for further information stored. As such each landmark anchors a distinct region. Landmarks in mental spatial representations are graded by their embodied or mediated experience, which may stem from their varying salient properties, or alternatively from the frequency of exposure, or type of activity of a person in the environment. However, there is no systematic research evidence available to date on cognitive salience. In addition the objects forming landmarks belong to categories of different spatial granularity. As we have seen, the hierarchical structure of configurations of landmarks in mental spatial representations provides for efficient heuristics in spatial reasoning, accepting even distortions in spatial reasoning. Other distortions arise between differently graded landmarks, or by the local density of landmarks—the amount of objects in between [83].

Besides organizing information about space, many landmarks appear to form the bridge between mental visual imagery and mental spatial representations since they have both properties, visual imagery and spatial anchor. In this dimension one might identify a third hierarchy, one of visual detail.

Finally, use of the mental spatial representation is highly context-sensitive. Only portions relevant are activated for working memory.

3.4 Externalization of Mental Spatial Representations

Cognitive scientists also study externalizations of mental spatial representations, or the ways how people communicate about space via an external medium. Communication covers spoken or written language, but it includes also drawing (from sketches to maps) and gesturing. Each of these modes gives clues of mental spatial representations and abilities, and in each of these modes we will encounter landmarks as a central concept. However, each of these modes also involves conscious (System 2) processes, and this fact lays another filter between the mental spatial representation and the observer.

As mental representations and language are seen as correlated in general [93, 168], language and space can be expected to be correlated as well. The ability to coordinate, and hence, communicate about locations and directions must be one of the most ingrained language abilities, as some explain by pointing to foraging and escaping [242]. Also, experience of space is fundamental for a body and its movement, and learned early (see Sect. 3.3.2.1). Hence it is no surprise that spatial language terms are mapped to other domains in metaphors [108, 112].

Despite a universal experience of the body in space, Levinson [120] explains that human spatial thinking is heavily influenced by culture. Different cultures can conceptualize space differently, which shows in various language structures. Also Mark et al. (e.g., [136, 137]) have demonstrated cultural differences between the conceptualizations of geographic forms.

In this respect, spatial language can only be investigated in a language-conscious manner; see for example the careful limitation to English in the work of Herskovits [81] and many others in the field, or Mark's assertion of differences between cross-cultural conceptualizations, and correspondingly, cross-linguistic terms [134].

Talmy [208] identified strong Gestalt properties in the spatial language of localizing an object. He found that in an utterance conveying spatial information the object to be localized, the *figure*, is typically smaller, or more movable or variable, or more salient. The figure is localized by references to known, larger, more complex and more stable objects, the *ground*. This way language structures space for the mental spatial representation of the recipient. Tversky and Lee responded to Talmy by pointing out that not only language structures space but also space structures language, this means, whichever language is chosen, graphical or verbal, similar kinds of information will be omitted or retained [223].

Language is also not precise, adding to the uncertainty of an already abstracting mental spatial representation (see Sect. 3.3.2). More complicated, the mental spatial representation of the speaker, or even only the subset in working memory, is not fully conveyed in language. Language is a different representation medium, hence, a transformation from mental spatial representation to language is required. This transformation is lossy. The mental spatial representation is configurational, and thus close to visual imagery, which is hard (or lengthy) to capture in all details in words, as the proverb knows: "A picture is worth 10,000 words" (see [115] for some history and scientific evidence). This argument for information loss relates also to the linearization that is part of the transformation. A configuration has two- or three-dimensional extent, and language is sequential. And last but not least, spatial language can further abstract from knowledge in a mental spatial representation. Language has been shown to be mostly qualitative about relations, while configurational knowledge may still represent quantities such as lengths or proportions. A qualitative description, however, can only invoke a schema or prototype in the recipient's mind, and quantities remain undefined [208]. Recipients even seem to reduce this indeterminacy by choosing preferential interpretations, specific interpretations that in the process of communication only get revised later if they lead to contradictions [105].

The recipient's mind, however, is not a blank sheet when receiving a (verbal or graphical) spatial description. Instead the recipient has prior spatial knowledge, at least of the procedural type. Recipients have now two tasks: matching a received description with their prior spatial knowledge, and using their constructed mental spatial representations for the intended purpose.

Communication is usually not a one-way activity. Klein, for example, has studied the dialog between two people about a route. The recipient can come back to the speaker, ask for more detail, or for confirmation [100]. Similarly, in an experiment called *Map Task* [7] participants had a dialogue about a route (see gray box). While the map task itself aimed at supporting research in speech recognition, it is

Table 3.1 Spatial and temporal communication constraints [90]

	Synchronous	Asynchronous
Physical co-presence	Face-to-face (human-human), or real-time mobile services (human-computer)	Refrigerator notes, you-are-here maps, departure plans at bus stations
Telepresence	Telephone	Email, accessing a static web page

also a remarkable illustration how complex verbal dialog can be, which is just conveying the position of a line in a diagram. Some of this complexity raises actually from the differences between the two communicating partners' knowledge of the environment, which requires negotiations and adjustments of knowledge.

The Map Task. In 1991 the Human Communication Research Centre at the Universities of Edinburgh and Glasgow designed an experiment to collect a corpus of 128 dialogues about maps. In their experiment two participants would "sit opposite one another, and each has a map which the other cannot see. One speaker—designated the *instruction giver*—has a route marked on her map; the other speaker—the *instruction follower*—has no route. The speakers are told that their goal is to reproduce the Instruction Giver's route on the Instruction Follower's map. The maps are not identical and the speakers are told this explicitly at the beginning of their first session. It is, however, up to them to discover how the two maps differ. The maps were designed as line drawings with landmarks attached. The landmarks were labelled with names. A variable in the test was the degree of agreement between the maps, where differences between the maps could consist of absence of some landmarks and name changes of landmarks, which also was used to produce ambiguities by multiple occurrences of a name.

groups.inf.ed.ac.uk/maptask/, last visited 3/1/2014.

In face-to-face communication non-verbal forms of communication such as gestures, facial expression or body language can play a role as well. However, spatial communication is not limited to face-to-face communication. Applying a schema proposed by Janelle [90] four levels of spatiotemporal constraints can be distinguished in human communication in principle (Table 3.1).

In each of these communication situations, spatial knowledge can be shared in different communication modes. For example, in each situation a map can be shared. In face-to-face situations both instruction giver and instruction taker can lean over a map and discuss a route together. In asynchronous, co-located communication the instruction giver can leave a map on the table to be picked up by the instruction taker at a later point in time. In synchronous, non-co-located communication, an instruction giver can use a map when talking to an instruction taker—the map task situation from above. And in asynchronous, non-co-located communication an instruction giver can draw a route on a map and send by mail to the instruction taker. But an equally perfect example, using another communication mode, is pointing directions. In face-to-face situations a pointing gesture in a direction can be sufficient, while in asynchronous, co-located communication a street sign pointing in a direction does provide the equivalent. The street sign has been left behind by somebody, the road authority, for the purpose of guiding any passer-by. In contrast, a mobile location-based service can show an arrow on the screen of the smartphone that is reflecting an up-to-date knowledge of a central (non-co-located) database. Its frame of reference is established locally from the positioning and the compass sensors of the smartphone. The smartphone can also access websites that may present sketches indicating directions by arrows, which establishes an asynchronous and non-co-located form of communication. The shown sketches have to provide also a frame of reference, for example, a landmark with its landmark-centered reference system.

3.4.1   The Externalized Message

In this context most revealing about mental spatial representations is the message produced, which is focusing on the producer only. Collecting gestures, utterances or observing other communication behavior of an individual should enable to reconstruct aspects of their mental spatial representations. However, before we review the corresponding literature in the following sections let us reflect on this methodology and its implications. Two issues in particular shall be discussed: the levels in the externalization process, and the reading process of the scientist.

The prior relates to the construction and use of mental spatial representations. As discussed above, mental spatial representations of individuals are constantly evolving by their embodied or mediated experiences. Mental spatial representations have been understood so far as being in long-term memory (see Sect. 3.3.1). Particular tasks at hand invoke portions of the long-term memory into working memory, the mental visual imagery available for problem solving which also applies some spatial abilities. Then selected elements of the working memory—the portion that has been found relevant for the solution of the problem—has to be mapped into language (for this triad see, for example, [39]), such as signs or gestures, sketches or words. Each of these communication modes is flexible, such that there is no one-to-one correspondence of working memory and expression. Consider for example the large variety of ways of describing one and the same route, or one and the same place [187]. In a communication process only one of the many possible expressions will be realized. The point is that an expression in a language is a very indirect clue on mental spatial representations.

To illustrate this argument further, consider a local being approached by a passer-by: "Excuse me, do you know where the train station is?" (resuming the example around Fig. 1.7). The straight answer would be "Yes" but such a literal understanding of the question is clearly inappropriate considering the context of the question. What actually happens in the mind of the local, interrupted from some meditations on the way home, is switching to a problem solving mode and activating from long-term spatial memory what is needed to think about an answer for this question. *Train station* provides a clue, and its ambiguity needs to be resolved. The location of the encounter provides a clue, the fact that the passer-by is walking adds to context, and also the guess that she is unfamiliar with the environment. Certain memories pop up in the local's mind, candidate solutions, and he might search for a route simple to explain and memorize, and suited for walking. This all happens in a blink of an eye because he also has already started talking to show his preparedness to help: "Of course! Let me see …". After he has described the route to her and the two have departed he still thinks about this encounter, now in a more relaxed mood. Suddenly he remembers a landmark along the route he should have referred to for easing her wayfinding task. This late intuition must have been in his long-term memory all the time, but it could not be immediately activated. Also, with the benefit of hindsight he may even remember a more pleasant route. And then there comes back this particular sweet memory for one place along the route, which he did not mention at all but which makes him dream away now. To give this story its twist, imagine that the passer-by actually was a researcher testing the local's knowledge, or mental spatial representation of the locality. Certainly she received valuable insights, but she must also be aware that she never will have a full or even accurate picture.

The story is not yet complete, and allows to illustrate the reading process of the passer-by. While the local walks parts of this route he just described to the passer-by he notices a narrow laneway off the street he had not considered, even forgotten when he said: "At the second intersection turn left". He starts worrying whether the passer-by really has found her way when counting intersections. His worries refer not to the completeness of his mental spatial representation, but to the shared experience and conceptualization of reality, and the general indeterminacy of language. In this particular instance the problem is the meaning of *intersection*. This indeterminacy has been studied in speech act theory [9, 192, 239] and reading theory [47, 53]. Key in both theories are the intentions of the speaker (*intentio autoris*) and of the reader (*intentio lectoris*).

Eco points out that there is even a third, the *intentio opere* [47]. In our example, the intention of the local was to help another person finding their way. Conversational maxims were applied to produce an utterance (a spoken or written instruction, a sketch, a gesture) serving this purpose. Once words have been spoken or sketches drawn, they have a meaning by themselves, which, depending on the instruction giver's language skills, mood and care, will be close to his intention but never accurate, just for the mentioned indeterminacy in language. This (imperfect) transfer of a speaker's intention to an utterance is crucial especially in asynchronous communication contexts, where an utterance (say, a street sign, or a mobile map) will be read and interpreted in absence of the speaker, or where the speaker is unrecognizable or irrelevant [235]. In addition, the communication skills of the individual impact on the form and content of the message [226].

While this *intentio opere* is identical for any reader, the *intentio lectoris* will be individually different. The passer-by as a wayfinder seeks to match the meaning of words with the environment found in order to progress towards the train station, and as a researcher she seeks to isolate references to objects, and the spatial relationships between the objects in order to reconstruct a mental spatial representation. It is therefore likely that the wayfinder and the researcher interpret this utterance differently. For example, "turn left" will be interpreted by a wayfinder in situ, experiencing a complex street intersection with a variety of affordances and opportunities to move through, and the wayfinder will choose a way coming satisfactorily close to the prototypical left-turn at the expected location. The researcher, in contrast, will infer that the local knows a route involving two street segments and a particular direction relationship. The wayfinder's interpretation reveals itself in action, while the researcher's interpretation remains in the realm of language (probably a formal language such as first order logic). Both readers will fill indeterminacies in the instruction according to their reading context, applying the basic logic of cognitive models [111].

A further distinction regarding the cognitive economy in communication applies to the way how a single landmark is referenced in an externalization. This distinction is grounded in base level theory. Base level theory [180] postulates that taxonomies of categorizations show categories of optimal abstraction at certain levels. These base categories are those of maximal category resemblance (category resemblance was introduced in Sect. 1.1.2). Base categories are typically between a superordinate level category that is more distinctive but also more abstract— too abstract to show strong resemblance, and a subordinate level that is more informative, but only slightly so and thus in many cases of unnecessary detail. For example, *furniture* is an abstract category and difficult to visualize, while its instance *chair* has strong resemblance linked to a visual prototype, while the subordinate category *stool* is more specific, but so specific that in many cases the reference "chair" is sufficient, and preferred.

Lloyd et al. [124] have shown that geographic categories are also organized according to base level theory. Geographic base level categories (country, region, state, city neighborhood) were associated with more common attributes than with the superordinate and more abstract category of *place* (Sect. 1.2.1), and that not significantly more attributes were associated with the more informative categories at subordinate level (such as home country, home state). Let us translate these observations into the context of landmarks. With an abstract category *landmark* as superordinate concept, the two messages "Turn left at the church" and "Turn left at St. Francis" are pragmatically identical allowing a wayfinder to make the same decision (assuming the next church is St. Francis). The prior is referring to a base category with strong category resemblance linked to a visual prototype. The latter one is more informative (even a categorical "Turn right at the Catholic church" would have been), but the additional amount of information is irrelevant in this context, and hence, neglected by many speakers. The additional information is only relevant—and then not neglected by speakers—in contexts where the a reference to the base category would be ambiguous; say, where St. Francis would be not the first church encountered and finer distinctions have to be made.

In the following two subsections we will focus more narrowly on *sketches* and spoken or written *verbal* instructions. We will leave aside other forms of external-izations, although they also could provide insights to mental spatial representations in general, and landmarks in particular. Maps, for example, are rarely the output of a single person's mental spatial representation, but rich in documenting collective human landmark experience in an environment. Purely symbolic representations such as arrows are externalizing aspects of mental spatial representations and studied in this regard (e.g., [102]), however, they are poor about landmarks. Pointing, as we have mentioned before in the context of path integration, provides an important insight into mental spatial representations, however, it stays also poor with respect to landmark knowledge. We also leave aside more literary descriptions, for example those that try to capture the atmosphere or essence of an environment.

The following two subsections aim in particular to extract these kinds of knowledge about mental spatial representations:

- Elements of mental spatial representations.
  We are asking what the atoms of mental spatial representations are. Lynch, for example, identified elements of graphical representations of cities [128]. But Lynch's work was limited to one particular context, or level of spatial granularity. More work is needed in various contexts, and should include research on qualities of landmarks, or on *landmarkness*, since we learned above already that all of Lynch's elements have landmarkness.
- Structures of mental spatial representations.
  We are asking how these atoms are connected. Here the hierarchies by granularity (containment) and by salience (use of anchor points) play a role. But also research in identifying fundamental qualitative spatial relationships belongs here. Mark and Egenhofer, for example, have studied the cognitive meaning of spatial predicates in spoken languages by presenting participants labelled sketch maps and asking for agreement to the label [135].
- Individual landmarks.
  We are asking which objects in a specific environment have formed the experience and ensuing mental spatial representation of an individual. Lynch's sketches do not only reveal elements (categories of types), they also reveal concrete instances of these elements. In a similar vein, collecting landmarks from corpora of place descriptions [39] or from information retrieval approaches (e.g., [24, 176]) should allow a mapping of individual, collective, and context-aware mental spatial representations of particular environments, despite our awareness that completeness can never be reached.

3.4.2 Sketches

A sketch, in contrast to verbal language, is a pictorial representation. We consider here sketches describing spatial locations, spatial configurations, or routes through an environment. These sketches convey visually a subset of a mental spatial representation, filtered for relevance, abstracted, schematized and mostly qualitative. A prime example of sketches are subway maps, which are highly schematized, and adhering to network topology rather than geometry. People also sketch in everyday situations and with much less rigour to schemas, often quickly drawn on a napkin. Studying sketches has become a standard method in spatial cognition research to learn about the individual's mental spatial representation. Sketches produced by humans contain also all the cognitive distortions of mental spatial representations [52, 232]. The scientific approach of asking people to draw is acceptable since test-retest experiments have shown that people produce highly correlated sketches over time [10].

Already Lynch asked people to draw sketches of their home town [128]. Lynch was interested in the legibility of cities, or in the ease of conceptualizing the familiar layout of the city in internal mental spatial representations. Thus these sketches can be understood as direct externalizations of the participants' mental spatial representations, although focused on the communication context set out by Lynch's instructions. Lynch observed that these sketches contain mainly five types of elements. He identified *paths*, *edges*, *districts*, *nodes*, and *landmarks*. These are the (shortened) definitions provided by Lynch ([128], p. 47)—examples in italics added by us:

1. Nodes are points, the strategic spots in a city into which an observer can enter, and which are the intensive foci to and from which they are travelling. They may be primarily junctions, places of a break in transportation, a crossing or convergence of paths.
   Think of a busy street intersection or a popular place in the city center.
2. Landmarks are another type of point-reference, but in this case the observer does not enter within them, they are external.
   Think of a store, a monument, or a school.
3. Paths [are] channels along which the observer [...] moves. [...] they are the dominant elements [of a city] image. People observe the city while moving through it, and along these paths other environmental elements are arranged and related.
   Think of roads, trails, sidewalks, or underground passages.
4. Edges are the linear elements not used or considered as paths. They are boundaries between two phases, linear breaks in continuity.
   Think of walls, seashores, or railway embankments.
5. Districts [...] are sections of the city, conceived as having two-dimensional extent, which the observer mentally enters 'inside of', and which are recognizable as having some common, identifying character. Always identifiable from the inside, they are also used for exterior reference if visible from outside.
   Think of a wealthy neighborhood, a business district, or a nightlife quarter.

Strikingly these elements can be grouped into two groups [217]. Paths, nodes and districts facilitate or afford *movement* of people. The second group is formed by elements that inhibit movement, namely edges and landmarks. These elements either increase the integration of the environment and contribute to its cohesion and homogeneity, for example, paths connecting parts of the city, nodes connecting paths, and districts formed by perceiving groups of nodes, paths and landmarks as regions of homogeneous character. Or they increase its heterogeneity by fragmentation and differentiation, for example, edges separating or delineating districts, and landmarks representing islands of distinction in a city.

Thus Lynch's classification, despite its impact on our understanding on mental spatial representations, seems to be focused to the context of wayfinding and orientation at urban scale. Stevens [205] studied playful behaviors in urban contexts and replaced some of Lynch's elements by elements of more private nature. He introduces *thresholds* and *props*. Thresholds are the locations where paths cross boundaries, such as the entrance to the train station, or the stairs up to the library, where juveniles or street artists find niches. Props are elements of 'urban detail', small and easily overlooked but perhaps producing a private experience for some, such as public artworks, signs, trees, street furniture, or doorknobs. Such observations clearly support the expectation that elements are context-dependent, and vary over scales of granularity.

Another extension of Lynch's elements is also relevant for our exploration. In Lynch's classification, a railway embankment is an edge, since pedestrians and motorists have to travel detours to find crossings. But switching the context, and considering public transport users instead, a train line becomes a path. This means that Lynch's elements, when applied to objects in the city, depend on a person's perspective. This perspective relates the person's mobility characteristics with the environmental affordances [217].

A special affordance in this regard is accessibility, since already Lynch had distinguished between accessible elements (path, node, districts) and inaccessible elements (edges and landmarks). Following logical deduction one additional element has to be added [217]. Lynch's accessible elements are zero-dimensional (node), one-dimensional (path), and two-dimensional (district), while his inaccessible elements are zero-dimensional (landmarks as reference points) and one-dimensional (edges). There must exist a two-dimensional inaccessible element, a *restricted district*, such as barracks, waste land or gaps that are inaccessible for (civilian) pedestrians, motorists, or other mobility modes. The reason why neither Lynch nor anybody else has postulated it before is probably founded in the sketches themselves. Restricted areas do not get sketched. They are the blank spaces on sketches. However, keeping in mind that sketches are drawn for a purpose, if the person switches their perspective the blank spaces in one context can become visible in another context.

Thus, Lynch's elements reflect the sketching person. They represent the relevant objects of a mental spatial representation for wayfinding and orientation assuming a specific mobility characteristic. Different perspectives can produce different uses of these elements. With other words, as externalizations of mental spatial representation these sketches are producing only views.

Lynch's *landmark* is of course substantially different from our own notion as laid out in Sect. 1.1.3. Nowadays it is generally acknowledged that all of Lynch's elements can have some *landmarkness* in terms of cognitive salience. Milgram, supported by Jodelet, collected a similar dataset to Lynch's, for the city of Paris. His *psychological maps of Paris* [143] reveal not only the imperfect mapping between reality (the city in stone) and people's mental spatial representations, as we would now expect after the discussion above. They also reveal the individual reference points people have used in their sketches. Milgram is aware of this additional outcome: He lists the "50 most frequently cited elements", a list of landmarks that—not surprisingly for those who have ever visited Paris—starts with Seine, l'Étoile, Tour Eiffel, Notre Dame, and Champs-Élysées. Already these few top elements from the list are supporting the argument that all of Lynch's elements can have some landmarkness. The Seine would be a barrier (for pedestrians and cars) according to Lynch, l'Étoile a node, and Champs-Élysées a path.

3.4.3 Verbal Descriptions

People also externalize their mental spatial representations when they give verbal, i.e., spoken or written descriptions about locations, configurations, or directions to a recipient. Similar to sketches, landmarks are reference points for anchoring verbal descriptions in location.

A fundamental observation about landmarks in verbal descriptions has been made by Landau and Jackendoff [114]. They observed that the located objects (locatum) appear to be encoded in language with more detailed geometric properties such as their axis, their volume, surfaces and parts. In contrast, reference objects (relata) are encoded only with coarse geometric properties, primarily the main axes. Their observation is consistent with our expectation that the relata are shared knowledge. They do not require description except what is needed to establish the frame of reference for the locatum, i.e., information about their orientation. In addition, they found the preference for prepositions of qualitative spatial relationships, especially topology, distance and direction. They conclude: "The striking differences in the way language encodes objects [locata] versus places [relata] lead us to [postulate] a non-linguistic disparity between the representations of *what* and *where*", a disparity that already had been discussed above to be detected in the visual apparatus [13,191], as well as in the neuronal basis of the mental spatial representation [104]. In a cross-linguistic comparison between English, Japanese and Korean, Munnich et al. added that spatial properties show sufficient similarity between languages to assume a common cognitive basis [153].

These verbal descriptions cover the same communication purposes as the sketches discussed above, but use a different language, namely a non-visual and linear language. Verbal descriptions are sequential in utterance and understanding.

As this linear structure will have less impact on route descriptions, since routes themselves are linear, most research on verbal descriptions so far has looked into route descriptions. Couclelis [30], for example, pointed out the potential to learn about spatial cognition through the investigation of verbal route descriptions. She wrote:

> Route directions are readily available, natural protocols reflecting the direction givers' cognitive representations of certain critical aspects of their environment. Still, the relationship between the spoken words and the underlying cognitive structures is far from transparent. Responding effectively to a non-trivial request for route directions is a complex task during which different aspects of spatial cognition come into play at different stages. A number of questions can be raised about that process (p. 133).

Some of the questions we aim to illuminate here.

Yet the linear structure of language will have a stronger impact on location and configuration descriptions. These descriptions require cognitive strategies to linearize a mental visual image during the production of the description, and to recombine to a image by the recipient. Thus, in the following we distinguish especially *route descriptions* [100], guiding through an environment or to a particular destination, and *place descriptions*, answering *where* questions for objects or configurations of objects [194]. For both kinds of research collections of such descriptions are required, *text corpora*. However, we are not aware of any test-retest experiments clarifying whether people produce over time correlated verbal route or place descriptions, as had been shown for sketches [10].

3.4.3.1 Route Descriptions

Route descriptions are essential for sharing spatial knowledge or coordinating collaboration. We will find that route descriptions are a prime case for the use of landmarks. Actually, the overwhelming majority of human route descriptions is preferring references to landmarks over geometric descriptions [40, 142]. Thus, a route instruction "At the next intersection turn right" is far more likely than "Turn right after 121 m", even if both descriptions are pragmatically equivalent enabling an instruction follower to turn at an indicated location [54].

To facilitate a comparison of different route instructions Frank [54] suggested a concept of *pragmatic information content*, a term not to be confused with the well-known entropy-based information content of Shannon and Weaver [195]. He writes: "A theory for a measure of pragmatic information content must account for the fact that different messages may have the same content"—route instructions of different people lead along the same route—"and that the same message may have different content for different recipients"—route instructions for a pedestrian may be hard to follow by a car driver.

(continued)

(continued)

"Information is only useful pragmatically when it influences a decision", and thus, pragmatic information content can only be determined for a particular user in a particular decision making situation. "All messages which lead to the same actions have the same information content, which is the minimum [size of the instruction] to determine the action [for this user]. If two users differ in the action they consider, their [situations] differ and therefore the information they deduce from the information content of the same message is different" (p. 47).

A route description is revealing a mental spatial representation in two ways. One insight is given by the actual route chosen, among all possible. Yet the only valid conclusion is that this route is known by the speaker, and no inference can be drawn on the set of alternative routes available to this speaker. The other insight to the speaker's mental spatial representation is the structure and richness in which this route is described. Which brings up the issue of the quality of route descriptions. What are the characteristics of a *good* route description given that descriptions are context dependent? More specifically, is a richer route description a better route description? Or is at least the person with the richer mental spatial representation producing better route descriptions? It should become clear from these questions that route descriptions can only provide a limited insight into the content of mental spatial representations, and they reveal more about the spatial and verbal abilities of the speaker working with mental spatial representations.

The hint on pragmatic information content should help clarifying a notion of quality. The formal measure of pragmatic information content is based on the principles of brevity and relevance. Both principles were among Grice's *conversational maxims* for formulating a message: be brief, be relevant. Similarly, Sperber and Wilson in their *relevance theory* assume that the recipient of the message searches first for relevant interpretations, and does not search further for alternative meanings as long as no conflicts require a revision of beliefs [202]. Even a cognitive motivation for these principles can be given by referring to the capacity of the human short-term memory, as Clark did [27]:

> In general, evolved creatures will neither store nor process information in costly ways when they can use the structure of the environment and their operations upon it as a convenient stand-in for the information-processing operations concerned. That is, know only as much as you need to know to get the job done. (p. 64)

Based on these two principles it can be stated that all route descriptions enabling an instruction follower to realize a certain route are equivalent from a pragmatic perspective. This means:

- There will be longer (richer) and shorter (leaner) route descriptions facilitating an instruction follower to reach a target. The maxim of brevity would prefer the shorter ones. In the longer ones the maxim of relevance will identify irrelevant or redundant references.
- There will be route descriptions that will fail to guide a particular instruction follower to a target. These descriptions violate the maxim of relevance by omitting relevant information.

Thus, a different instruction follower may require a different message. However, one important conclusion follows from these rather abstract considerations. Of all pragmatically equivalent descriptions guiding to the target the shorter descriptions may be the better route descriptions. A tangible reason for this assumption is the limited capacity of short term memory [32, 110, 144]. Thus, in accord with other forms of externalizations, route descriptions reveal more about the structure of mental spatial representations in connection with strategies of spatial and communication abilities, and less about the content of mental spatial representations.

The assumption of preferences for shorter descriptions has actually been confirmed in independent research offering further insight in the internal structure and content of *good* route descriptions (e.g., [38–40, 127]). This cognitive and linguistic research has again to rely on indirect observations since there are no formal criteria for judging the quality of route descriptions other than whether the instruction taker has reached the target. Indirect ways of observation are:

- A purely descriptive approach of linguistic structure. The result is a characterization of a route description rather than an assessment. It permits at least a qualitative comparison between route descriptions for the same route.
- Ratings of human route descriptions by local experts. In principle this method can be applied in situ, e.g., after route following, or in a survey relying on the mental spatial representations of the raters.
- Navigational performance by instruction followers unfamiliar with the environment. In principle this method can collect whether followers succeed, but in addition can also survey how comfortable they felt.
- Comparison of human route descriptions with some algebraically produced route descriptions. Algebraic approaches are suited to produce minimal instructions according to some model, but since the model can mismatch with a context there is no guarantee for producing successful descriptions, let alone shortest successful ones. Hence, generally the algebraically produced route description have to be tested in a control experiment as well.

One of the first investigations of this kind were Wunderlich and Reinelt's [241] linguistic study of forms of speech in route descriptions. They worked from a corpus of route descriptions to identify four phases in the full discourse: an opening ("Excuse me, can you tell me ..."), the route instructions itself (the path to be followed), an optional securing phase (ensuring that the message has been conveyed), and a closure ("Thank you"). In the route instructions they found patterns identifying landmarks as intermediate destinations and locations of reorientation. This structure was close to a formal model proposed before by Kuipers [109].

Later Streeter et al. [207] applied a pure navigational performance test to compare customized route maps with verbal route descriptions. They demonstrated that car drivers following the verbal route descriptions, which provided one instruction per turn, drove fewer miles, took less time, and showed about 70 % fewer errors than the drivers relying on the route map. If a picture tells more than 1,000 words then less is obviously more. Correspondingly, drivers who had both verbal route descriptions and the route maps available performed badly as well.

Then Denis and collaborators [37–39] collected campus route descriptions from students. They applied all four approaches in their study of these route descriptions. In later work [40] Denis et al. repeated the experiment in the real world, collecting route descriptions from citizens of the city of Venice, and confirmed the prior findings.

First, they characterized the collected route descriptions by criteria such as actions specified and localized by a landmark, actions specified without a reference to a landmark, references to a landmark without any specified action, descriptions of a landmark, and comments. Results documented that people with higher visuo-spatial imagination also use more landmarks in their descriptions, which supports our claim that landmarks bridge between visual memory and spatial memory.

Secondly, these instructions were rated. The rating was performed by local experts and non-experts.

In addition, and related to the fourth approach, they came up with a construction of minimal descriptions, which they call *skeletal* descriptions. For the construction they used the elements identified in the student generated route descriptions, thus maintaining the perspective taking of the speakers whilst concentrating on the smallest common denominator in the descriptions.

And fourth, they tested navigational performance of instruction followers equipped with good, poor, or skeletal route descriptions.

Three of their observations are critical for us:

- The construction of skeletal descriptions confirmed that "landmarks and their associated actions were key components of [good] route description" ([39], p. 409). Also, references to landmarks are unevenly distributed along the route. They tend to concentrate at points where orientation decisions are to be made.
- Rating of the original route descriptions is highly correlated with their similarity to the skeletal description.
- Skeletal descriptions received scores similar to those of good descriptions, despite being, on average, shorter by design.

Independently, Lovelace et al. [127] collected a corpus of route descriptions for a particular route, and then searched for shared characteristics such as numbers of segments or turns, or numbers of references to landmarks used. Their classification schema was inspired by Denis' et al. but further refined by grouping landmarks in those at decision points and those not at decision points. While Denis et al. report of a tendency for concentration of landmark references at decision points, Lovelace et al. find about half of the landmark references not at decision points. Considering that both types of landmark references serve different purposes—the first one is anchoring an action of orientation decision making, while the second one is not linked to any decision and thus rather of calming or confirming nature—the different observations between the studies can relate to different contexts. Environments with longer route segments suggest intermediate confirmatory comments, especially for instruction followers unfamiliar with the environment.

This thought matches another observation of Lovelace et al. They asked people familiar and people unfamiliar with the environment to rate the collected route descriptions. Again in contrast to Denis' et al. findings of high ratings and high navigational performance with shorter route descriptions, Lovelace et al. report a preference for richer route descriptions (inter-rater correlations showed that subjective ratings were reliable and consistent across individuals). This preference for richer descriptions indicates again that within their context intermediate confirmatory references to landmarks were advisable.

Allen [4] wrote about the findings of Denis et al. on the essence of good route descriptions: "The next step in this strategy may involve a formal description of the structure and components of these skeletal descriptions, which consist of a combination of directives and descriptives as described previously" (p. 335). This is, of course, what the rest of this book is about. But before we move on let us have a look at Allen's own work. He demonstrated that route descriptions are better remembered and lead to higher navigational performance when the production of the route descriptions build in few psycho-linguistic principles:

- The principle of spatio-temporal order: The spatial and temporal order of localizations in route descriptions should be consistent with the order in which these locations are experienced when traveling along the described route.
- The principle of referential determinacy: References to landmarks at points where decisions about orientation have to be made, and links to the proper action with the experience of the landmark.
- The principle of mutual knowledge: Delimiters describing topological, directional and distance relations in route descriptions are chosen according to the communication context, i.e., appropriate for the environment and for the instruction follower.

However, all research cited so far studied route descriptions in a homogeneous context, which typically assumes a person unfamiliar with the environment, and in a mono-modal movement, either walking or driving a car. They emphasize the role of decision points (for orientation) along otherwise linear route elements. One could expect then Allen's principles be satisfied by route descriptions of single granularity, which means one instruction per decision point. However, even in these circumstances human route descriptions are not necessarily of a constant granularity. Instead, elements are often grouped together (a process in memory and language sometimes called *chunking* [35, 103]). For example, an instruction "at the third intersection turn right" applies numerical chunking, and "follow the signs to the airport" produces a group replacing intermediate decision points by a procedural description (remember Clark's "know only as much as you need to know to get the job done"?).

The effect is even more dominant in route descriptions for travels through partly familiar environments or when using multiple modes of movement. An example for the prior is the route description to a conference: "At Beijing Airport, take the capital airport express train ...", which is completely ignoring the traveller's route to Beijing Airport. Adhering to above principles, the speaker is rightly assuming that the experienced conference traveller will manage this part on his or her own. Similarly, while transfers along multi-modal routes are relatively small scale in space and time, they put more responsibility on the wayfinder than the longer legs of the journey. Accordingly, descriptions of multi-modal routes adhering to above principles will vary in spatial granularity, providing information of finer spatial granularity for transfers [82,213].

3.4.3.2 Place Descriptions

Place descriptions are so common that we do not think much about them. A person tells her partner where the keys have been left, or where they should meet in the evening. They call a local emergency number when they have witnessed an accident and explain where this has happened. They write captions revealing locations when uploading holiday pictures to a social networking site. Social conventions even create new forms of place descriptions, such as *checking in* on one social networking site, or *hashtagging* a location on another site. All these conversations are performed with the intention to help the recipient identifying or finding these places. More complex place descriptions can describe whole configurations of objects. These descriptions are intending to help the recipient to form a mental image of an environment. For example, a person moving into a new apartment may send a letter to an old friend describing this apartment as a configuration of rooms. Or a second year student may explain the configuration of buildings on campus to a fresh first year student.

Both kinds of place descriptions are challenged by the linear structure of language. A description of a location has to refer to objects in the environment that are in some two-dimensional relationship with the location to be specified ("the café in Richmond"), or a three-dimensional relationship ("the key is on the living room table, below the newspaper"), or even a relationship that realizes a temporal dimension ("in front of the place where the café has been"). Now we already recognize that all references to relata are references to landmarks, and assumed to be known or recognizable by the recipient.

We call the located object the *locatum*. This object is located in relationship to one or several known objects, the *relata*. The relata form the frame of reference to locate the locatum. The speaker must assume knowledge of the relata to be shared by the recipient, such that the recipient can re-establish the frame of reference in their mind.

In English language, a common schema connecting the locatum and relatum looks like [228]:

$$< \texttt{locatum} \rightarrow \texttt{spatial\_relationship} \rightarrow \texttt{relatum} >$$

For example, in "Cartier is at place Vendôme" the jeweller shop *Cartier* is the locatum and *place Vendôme* is the relatum. Variations of this form exist, of course. For example, I may describe my location with "at home", which at the surface omits the locatum. Also, the schema describes only binary spatial relationships between one locatum and one relatum, but there are also ternary (or *n*-ary) spatial relationships between one locatum and two or more relata. "The fountain is between church and city hall" is an example for a ternary relationship.

Human place descriptions almost exclusively use qualitative spatial or temporal relationships to link a locatum with its relatum. While graphic languages (maps, sketches) still convey some geometric meaning, verbal place descriptions would know order ("behind the library") and even comparison ("a larger building"), but rarely metric information. Typically it is "I am close to the intersection" rather than "I am 30 m from the intersection", and "to the right" instead of "in 85°".

Several linguistic strategies have been identified helping with the linearization challenge of place descriptions (e.g., [37]). One of them is a deliberate choice of the speaker of a survey or a route perspective. In a survey perspective references are put in the sequence in which a sketch would be drawn ("find the café north of the library"). In a route perspective the linear sequence follows from mentally following a route ("find the café after passing the library"). In both cases landmarks (the library) help with localization. Hidden behind this observation is already a glimpse of an alternative linguistic strategy for linearization, which is zooming through hierarchies [169, 170, 194]. "The café is behind the library" is linking a less salient object (café) with a more salient object (library), i.e., zooming out on the salience hierarchy.

Both hierarchies, by salience as well as by spatial granularity, are suited to zoom in or zoom out. A sentence like "Cartier is in Paris, at place Vendôme" is zooming in through a hierarchy of spatial granularity, from the coarser Paris to the finer place Vendôme. The interpretation is actually complex:

1. Cartier is in Paris.
2. Cartier is at place Vendôme.
3. Context suggests that this is the place Vendôme in Paris.

The sentence could also have been "Cartier is at place Vendôme in Paris", zooming out. Grammatically, however, the interpretation of relata in this sentence remains ambiguous. It is either of both:

1. Cartier is at place Vendôme, which is in Paris.
2. Cartier is in Paris, and also at place Vendôme.

In the prior case the locatum is related to one relatum only, which in turn is related to another relatum. In the latter case both reference objects are related to the locatum.

In the prior case the spatial relationship between *place Vendôme* and *Paris* is given, in the latter it must be inferred. Similarly, anaphora can introduce ambiguity. The utterance "Cartier is at place Vendôme; it is in Paris" requires a resolution whether "it" refers to Cartier or place Vendôme. Western postal addresses are such place descriptions, applying the hierarchical pattern: "Cartier is at place Vendôme, and place Vendôme is Paris". The relatum at each level is located (or disambiguated for localization) by the next coarser level.

But why did the instruction giver found it necessary to add *in Paris* as a coarser relatum to an already finer localization? After all, adding relata in a zoom-out fashion cannot contribute to improving the localization precision like a zoom-in structure does. However, it can improve the accuracy. An instruction giver does not stop after *at place Vendôme* because she or he considers the utterance insufficient for one of two reasons. One is that the instruction giver is not sure whether this finer relatum is salient enough. In this case the better known *Paris* can trigger the memory for a less known *place Vendôme* via a salience hierarchy. The other possible reason is that the finer relatum is ambiguous. In this case *Paris* is used to disambiguate the place Vendôme in Paris from any other place Vendôme elsewhere.

It has not yet been investigated how the two hierarchical organization principles—cognitive salience and spatial granularity—are applied together in language production [178]. For example, "Cartier is next to the Ritz" would have been another valid place description, this time linking the location of the jeweller to the presumably more well-known location of the Hotel Ritz. Both are institutions characterized at the same level of spatial granularity, street address level. However, one is considered more salient than the other, or used as an anchor point to locate the other. Perhaps the speaker knows that the instruction giver had the experience of a stay at the Ritz before. Whether in a particular communication situation the speaker prefers to provide a hierarchical instruction by granularity or salience, is part of the flexibility of language.

3.4.4   Understanding External Spatial Representations

Mental spatial representations are not only formed by being exposed to an environment, but also by being exposed to secondary sources. All external representations discussed above contribute to the formation of a mental spatial representation, including printed maps or street signs.

The reading process of this mediated, second-hand information has many parallels with the reading process during the first-hand, in-situ experience of the environment. Neuroscientific research even supports that reading is an embodied process, in the sense that comprehension involves simulations of an embodied experience involving a reactivation of the reader's perceptual, motor and affective knowledge [26]. Thus, the in-situ experience is a reading process as well, by comprehension of a scene consisting of perception of properties and affordance, and an effort of the mind to make sense of it. Mediated information affords to be integrated in a mental spatial representation, especially in a problem solving process. In-situ one may experience an object standing out in the environment, a figure on a ground. Correspondingly, reading a map symbol or graphical structure on the ground of the paper, or listening to an route description that emphasizes a particular object for anchoring a left-turn, provides a similar figure-ground experience. Map objects stand out for a purpose, i.e., by intention of the cartographer. The real-world object represented by this medium is supposed to be relevant since it survived cartographic generalization, and it anchors spatial configuration. The same can be said for the object represented on a sketch, only that the sketch is even more focused on the current communication context and may provide even more trusted relevant references. The object referred to in a verbal route description, in the expectation of the reader, must be standing out in the environment, must be recognizable, or can even be matched with knowledge of the reader. The reader will expect landmarkness of these objects in the environment that matches the landmarkness experience in the reading process.

Research on understanding external spatial representations investigates their impact on mental spatial representations. In early work Taylor and Tversky, for example, asked participants in an experiment to read verbal descriptions of an environment, provided in either a route or a survey perspective [212]. They did not find differences in an ability to answer verbatim or inference questions, suggesting that the participants formed the same spatial mental models capturing the relations between landmarks from the descriptions. Also, when participants in another experiment studied maps and then had to draw them and to verbally describe them, and the order of doing so was studied, no difference in the organization was discovered, again suggesting the same mental spatial representation—which evidently also showed a hierarchical structure [211]. There are, of course, individual differences of cognitive styles, as demonstrated by Pazzaglia [164]. More recently, Lee and Tversky [116] have studied the comprehension times of the participants being presented route and survey perspective descriptions of an environment, enriched with references to landmark objects either of visual detail or of factual detail. They observed that landmarks are neutral to the perspective: the landmark descriptions did not increase the comprehension time, neither in route nor in survey perspectives, but visual landmark descriptions seemed to support the comprehension of descriptions with perspective switches.

It goes without saying that the way how people understand maps, sketches or verbal descriptions is relevant for generating these messages. Hence, more recently research moved into studying the reading of machine-generated spatial information, where cognitive science meets human-computer interaction design. However, cognitive research goes beyond usual methods of interaction research, such as user satisfaction measures. Instead, the machine-generated information is tested for comprehensibility and task success. Generally this research is only indirectly about understanding landmarks, but rather on human spatial conceptualizations of relationships (e.g., [22, 102, 117]) and on relevance and communication efficiency (e.g., [33, 189, 190]). This type of research will be elaborated in later chapters, in more appropriate contexts.

3.4.5   What We Have Learned About Landmarks
from Externalizations of Mental Spatial Representations

All externalizations of mental spatial representations, such as maps, sketches, and verbal descriptions, show rich use of landmarks. Triggered by at least two filtering processes—first the selection, aggregation and abstraction process related to maintaining or recalling the mental spatial representation of the speaker, and secondly the filtering by the transformation into the language of the communication medium—the externalizations tend to show what is relevant in a given communication context, and neglect the rest. For the same reasons externalizations tend to be context-dependent and are by no means a representative image of the mental spatial representation.

Landmarks provide the spatial reference frame for locating other objects in an environment. Means for locating are spatial relationships, such as distances, directions or orientations, projections, or containment. While graphical languages preserve some ability to represent these relationships in a quantitative manner, verbal descriptions, written or spoken, show a strong preference for qualitative expressions. We remember Lynch's observation that people familiar with a city show a tendency to rely more on landmarks in their sketches, which means they rely on uniqueness, distinctness or local contrast rather than continuity. For route directions, Michon and Denis [142] concluded that these references to landmarks are intended to facilitate the construction of a mental spatial representation by the recipient.

3.5   Summary

In summary, landmarks have a prominent role in spatial cognition, more specifically, as points of reference in forming, maintaining, analysing and communicating spatial mental representations. People equipped with this spatial mental representation interact with their environments that are the subject of their mental spatial representations, but also with mediated experiences of these environments through external spatial representations.

This complex system of physical environment, persons perceiving and interacting with the environment, and (more and more electronic) information about this environment has a classical form in the semiotic triangle [157] between a referent (the object in the environment), the reference (a word or symbol for the object), and an idea (a thought or knowledge about the object). Kuhn adopted the triangle already to explore the interplay between language and landscape [107], and our version is shown in Fig. 3.9. The 'knowledge in the mind' consists of the mental spatial representation of the world, which is descriptive (the landmark, route or survey knowledge described above), and strategies to utilize this representation, the heuristics. The 'knowledge in the world' is the shape of the environment as

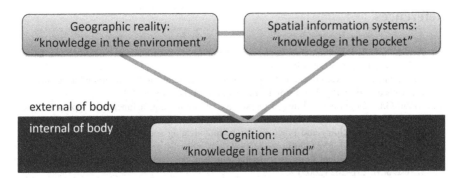

Fig. 3.9 Types of spatial and temporal knowledge involved in orientation and wayfinding

experienced, including objects standing out, signs or displays, announcements, or designs that afford particular behavior [172]. This knowledge is directly accessible to a person only within the range for perception, and it is typically stationary, but not necessarily static. Finally, the 'knowledge in the pocket' is the metaphor for all external representations of the environment, such as maps, sketches, or verbal descriptions. More and more these representations are digital, which are accessible for the person through a multitude of devices, and many of them portable or at arm's length for ubiquitous access.

The triangle should not hide the fact that the three nodes of the triangle are usually out of sync. The mental spatial representation has been formed by experiences over time, and each new encounter with a learned environment requires an ability to maintain this representation. Other sources for information, whether human or electronic, have also historic knowledge of the environment. These other accessible representations are also selective. The selection, aggregation and abstraction of objects to be stored were dependent on a particular context that is typically inaccessible in a current communication situation. Last but not least, representations are inherently uncertain. The latter is dictated by physics already. There is no such thing as an accurate observation of a reality such as an environment.

Interaction with the physical environment as well as with external representations of the environment is based on the concepts in a mental spatial representation, which, as we have seen, is heavily based on landmarks. Interaction with "knowledge in the pocket" requires the provider of the knowledge in the pocket to relate to the spatial mental representation for ease of use. People being able to pick up with ease presented information with their own mental spatial representation will be able to keep orientation and wayfinding at low cognitive load. This ability is important because both tasks are not primary purposes or activities of a human being. Movement always serves as a means to an end. Nonetheless, movement is an activity requiring constant interaction and coordination with the environment, including other people. Hence, "devices in our pockets" must become intelligent. The intelligent machine, in this sense, is a machine interacting with humans on their terms [55] when addressing spatial problems. If only for this reason the machine must understand landmarks for intelligent interaction with people.

References

1. Acredolo, L.P.: Developmental changes in the ability to coordinate perspectives of a large-scale space. Dev. Psychol. **13**(1), 1–8 (1977)
2. Allen, G.L.: A developmental perspective on the effects of "subdividing" macrospatial experience. J. Exp. Psychol. Hum. Learn. Mem. **7**(2), 120–132 (1981)
3. Allen, G.L.: Cognitive abilities in the service of wayfinding: a functional approach. Prof. Geogr. **51**(4), 554–561 (1999)
4. Allen, G.L.: Principles and practices for communicating route knowledge. Appl. Cognit. Psychol. **14**(4), 333–359 (2000)
5. Allen, G.L.: Functional families of spatial abilities: poor relations and rich prospects. Int. J. Test. **3**(3), 251–262 (2003)
6. Allen, G.L., Kirasic, K.C., Siegel, A.W., Herman, J.F.: Developmental issues in cognitive mapping: the selection and utilization of environmental landmarks. Child Dev. **50**, 1062–1070 (1979)
7. Anderson, A., Bader, M., Gurman Bard, E., Boyle, E., Doherty, G., Garrod, S., Isard, S., Kowtko, J., McAllister, J., Miller, J., Sotillo, C., Thompson, H., Weinert, R.: The HCRC map task corpus. Lang. Speech **34**(4), 351–366 (1991)
8. Appleyard, D.: Why buildings are known. Environ. Behav. **1**(2), 131–156 (1969)
9. Austin, J.L.: How to Do Things with Words. Clarendon Press, Oxford (1975)
10. Blades, M.: The reliability of data collected from sketch maps. J. Environ. Psychol. **10**(4), 327–339 (1990)
11. Blades, M.: The development of the abilities required to understand spatial representations. In: Mark, D.M., Frank, A.U. (eds.) Cognitive and Linguistic Aspects of Geographic Space. NATO ASI Series D: Behavioural and Social Sciences, vol. 63, pp. 81–116. Kluwer Academic Publishers, Dordrecht (1991)
12. Blades, M.: Wayfinding theory and research: the need for a new approach. In: Mark, D.M., Frank, A.U. (eds.) Cognitive and Linguistic Aspects of Geographic Space. NATO ASI Series D: Behavioural and Social Sciences, vol. 63, pp. 137–165. Kluwer Academic Publishers, Dordrecht (1991)
13. Braddick, O.J.: Computing 'what' and 'where' in the visual system. In: Eilan, N., McCarthy, R., Brewer, B. (eds.) Spatial Representation: Problems in Philosophy and Psychology, pp. 340–355. Basil Blackwell Ltd., Oxford (1993)
14. Briggs, R.: Urban cognitive distance. In: Downs, R.M., Stea, D. (eds.) Image and Environment, pp. 361–388. Aldine Publishing Company, Chicago (1973)
15. Brosset, D., Claramunt, C., Saux, E.: A location and action-based model for route descriptions. In: Fonseca, F., Rodriguez, M.A., Levashkin, S. (eds.) GeoSpatial Semantics, Lecture Notes in Computer Science, vol. 4853, pp. 146–159. Springer, Berlin (2007)
16. Brun, V.H., Solstad, T., Kjelstrup, K.B., Fyhn, M., Witter, M.P., Moser, E.I., Moser, M.B.: Progressive increase in grid scale from dorsal to ventral medial entorhinal cortex. Hippocampus **18**(12), 1200–1212 (2008)
17. Burgess, N.: Spatial memory: how egocentric and allocentric combine. Trends Cogn. Sci. **10**(12), 551–557 (2006)
18. Burgess, N., Maguire, E.A., O'Keefe, J.: The human hippocampus and spatial and episodic memory. Neuron **35**(4), 625–641 (2002)
19. Burnett, G., Smith, D., May, A.: Supporting the navigation task: characteristics of 'good' landmarks. In: Hanson, M.A. (ed.) Contemporary Ergonomics 2001, pp. 441–446. Taylor & Francis, London (2001)
20. Campbell, J.: The role of physical objects in spatial thinking. In: Eilan, N., McCarthy, R., Brewer, B. (eds.) Spatial Representation, pp. 65–95. Blackwell, Oxford (1993)
21. Carroll, J.B.: Human Cognitive Abilities: A Survey of Factor-Analytic Studies. Cambridge University Press, Cambridge (1993)

22. Casakin, H., Barkowsky, T., Klippel, A., Freksa, C.: Schematic maps as wayfinding aids. In: Freksa, C., Habel, C., Brauer, W., Wender, K.F. (eds.) Spatial Cognition II, Lecture Notes in Artificial Intelligence, vol. 1849, pp. 54–71. Springer, Berlin (2000)

23. Chan, E., Baumann, O., Bellgrove, M.A., Mattingley, J.B.: From objects to landmarks: the function of visual location information in spatial navigation. Front. Psychol. **3**, 304(11 pages) (2012)

24. Chen, W.C., Battestini, A., Gelfand, N., Setlur, V.: Visual summaries of popular landmarks from community photo collections. In: Xu, C., Steinbach, E., El Saddik, A., Zhou, M. (eds.) 17th ACM International Conference on Multimedia, pp. 789–792. ACM, Beijing (2009)

25. Cheng, K.: A purely geometric module in the rat's spatial representation. Cognition **23**(2), 149–178 (1986)

26. Chow, H.M., Mar, R.A., Xu, Y., Liu, S., Wagage, S., Braun, A.R.: Embodied comprehension of stories: interactions between language regions and modality-specific neural systems. J. Cogn. Neurosci. **26**(2), 279–295 (2014)

27. Clark, A.: Microcognition: Philosophy, Cognitive Science and Parallel Distributed Processing. MIT Press, Cambridge (1989)

28. Cohen, R. (ed.): The Development of Spatial Cognition. Lawrence Erlbaum Associates, Hillsdale (1985)

29. Cornell, E.H., Heth, C.D., Alberts, D.M.: Place recognition and way finding by children and adults. Mem. Cogn. **22**(6), 633–643 (1994)

30. Couclelis, H.: Verbal directions for way-finding: space, cognition, and language. In: Portugali, J. (ed.) The Construction of Cognitive Maps. GeoJournal Library, vol. 32, pp. 133–153. Kluwer, Dordrecht (1996)

31. Couclelis, H., Golledge, R.G., Gale, N., Tobler, W.: Exploring the anchorpoint hypothesis of spatial cognition. J. Environ. Psychol. **7**(2), 99–122 (1987)

32. Cowan, N.: The magical number 4 in short-term memory: a reconsideration of mental storage capacity. Behav. Brain Sci. **24**(1), 87–114 (2001)

33. Cuayáhuitl, H., Dethlefs, N., Richter, K.F., Tenbrink, T., Bateman, J.: A dialogue system for indoor way-finding using text-based natural language. Int. J. Comput. Linguist. Appl. **1**(1–2), 285–304 (2010)

34. Dabbs, J.M., Chang, E.L., Strong, R.A., Milun, R.: Spatial ability, navigation strategy, and geographic knowledge among men and women. Evol. Hum. Behav. **19**(2), 89–98 (1998)

35. Dallal, N.L., Meck, W.H.: Hierarchical structures: chunking by food type facilitates spatial memory. J. Exp. Psychol. Anim. Behav. Process. **16**(1), 69–84 (1990)

36. Damasio, A.: Self Comes to Mind: Constructing the Conscious Brain. Vintage Books, New York (2010)

37. Daniel, M.P., Carite, L., Denis, M.: Modes of linearization in the description of spatial configurations. In: Portugali, J. (ed.) The Construction of Cognitive Maps. GeoJournal Library, vol. 32, pp. 297–318. Kluwer Academic, Dordrecht (1996)

38. Daniel, M.P., Tom, A., Manghi, E., Denis, M.: Testing the value of route directions through navigational performance. Spatial Cognit. Comput. **3**(4), 269–289 (2003)

39. Denis, M.: The description of routes: a cognitive approach to the production of spatial discourse. Curr. Psychol. Cognit. **16**(4), 409–458 (1997)

40. Denis, M., Pazzaglia, F., Cornoldi, C., Bertolo, L.: Spatial discourse and navigation: an analysis of route directions in the city of Venice. Appl. Cognit. Psychol. **13**(2), 145–174 (1999)

41. Descartes, R.: Discours de la methode pour bien conduire sa raison, et chercher la verite dans les sciences. Plus La dioptrique. Les meteores. Et La geometrie. Qui sont des essais de cete methode. Ian Maire, Leyden (1637)

42. Dijkstra, E.W.: A note on two problems in connexion with graphs. Numer. Math. **1**, 269–271 (1959)

43. Downs, R.M., Stea, D.: Image and Environment. Aldine Publishing Company, Chicago (1973)

44. Downs, R.M., Stea, D.: Maps in Minds: Reflections on Cognitive Mapping. Harper and Row, New York (1977)
45. Duckham, M., Kulik, L.: "Simplest" paths: automated route selection for navigation. In: Kuhn, W., Worboys, M., Timpf, S. (eds.) Spatial Information Theory, Lecture Notes in Computer Science, vol. 2825, pp. 169–185. Springer, Berlin (2003)
46. Dudchenko, P.A.: Why People Get Lost: The Psychology and Neuroscience of Spatial Cognition. Oxford University Press, Oxford (2010)
47. Eco, U., Rorty, R., Culler, J., Brook-Rose, C.: Interpretation and Overinterpretation. Cambridge University Press, Cambridge, UK (1992)
48. Ehrenfels, C.v.: Über Gestaltqualitäten. Vierteljahresschrift für wissenschaftliche Philosophie **14**, 249–292 (1890)
49. Ekstrom, A.D., Kahana, M.J., Caplan, J.B., Fields, T.A., Isham, E.A., Newman, E.L., Fried, I.: Cellular networks underlying human spatial navigation. Nature **425**(6954), 184–188 (2003)
50. Etienne, A.S., Jeffery, K.J.: Path integration in mammals. Hippocampus **14**(2), 180–192 (2004)
51. Evans, G.W., Brennan, P.L., Skorpanich, M.A., Held, D.: Cognitive mapping and elderly adults: verbal and location memory for urban landmarks. J. Georontology **39**(4), 452–457 (1984)
52. Forbus, K.D., Usher, J., Lovett, A., Lockwood, K., Wetzel, J.: Cogsketch: sketch understanding for cognitive science research and for education. Top. Cognit. Sci. **3**(4), 648–666 (2011)
53. van Fraassen, B.C.: Literate experience: the [de-, re-] interpretation of nature. Versus **85/86/87**, 331–358 (2000)
54. Frank, A.U.: Pragmatic information content: how to measure the information in a route description. In: Duckham, M., Goodchild, M.F., Worboys, M. (eds.) Foundations in Geographic Information Science, pp. 47–68. Taylor & Francis, London (2003)
55. French, R.M.: Moving beyond the Turing test. Commun. ACM **55**(12), 74–77 (2012)
56. Freundschuh, S.M., Egenhofer, M.J.: Human conceptions of spaces: implications for geographic information systems. Trans. GIS **2**(4), 361–375 (1997)
57. Frisch, K.v.: The Dancing Bees: An Account of the Life and Senses of the Honey Bee. Methuen, London (1954)
58. Frisch, K.v.: The Dance Language and Orientation of Bees. Harvard University Press, Cambridge (1993)
59. Fyhn, M., Molden, S., Witter, M.P., Moser, E.I., Moser, M.B.: Spatial representation in the entorhinal cortex. Science **305**(5688), 1258–1264 (2004)
60. Gardner, H.: Frames of Mind: The Theory of Multiple Intelligences, EBL, vol. 3rd. Basic Books, New York (2011)
61. Gärling, T., Lindberg, E., Böök, A.: Cognitive mapping of large-scale environments: the interrelationship of action plans, acquisition, and orientation. Environ. Behav. **16**(1), 3–34 (1984)
62. Gärling, T., Böök, A., Lindberg, E.: Adults' memory representations of the spatial properties of their everyday physical environment. In: Cohen, R. (ed.) The Development of Spatial Cognition, pp. 141–184. Lawrence Erlbaum Associates, Hillsdale (1985)
63. Gibson, J.J.: The Ecological Approach to Visual Perception. Houghton Mifflin Company, Boston (1979)
64. Gigerenzer, G., Todd, P.M., Group, A.R. (eds.): Simple Heuristics That Make Us Smart. Evolution and Cognition. Oxford University Press, New York (1999)
65. Giocomo, L.M., Moser, M.B., Moser, E.I.: Computational models of grid cells. Neuron **71**(4), 589–603 (2011)
66. Golledge, R.G.: Place recognition and wayfinding: making sense of space. Geoforum **23**(2), 199–214 (1992)
67. Golledge, R.G.: Human wayfinding and cognitive maps. In: Golledge, R.G. (ed.) Wayfinding Behavior, pp. 5–45. The Johns Hopkins University Press, Baltimore (1999)
68. Golledge, R.G. (ed.): Wayfinding Behavior: Cognitive Mapping and Other Spatial Processes. The Johns Hopkins University Press, Baltimore (1999)

69. Golledge, R.G., Stimson, R.J.: Spatial Behavior: A Geographic Perspective. The Guildford Press, New York (1997)
70. Gopnik, A.: The theory theory as an alternative to the innateness hypothesis. In: Antony, L.M., Hornstein, N. (eds.) Chomsky and His Critics, pp. 238–254. Blackwell Publishing Ltd, New York (2003)
71. Gopnik, A.: Causality. In: Zelazo, P.D. (ed.) The Oxford Handbook of Developmental Psychology, vol. 1. Oxford University Press, Oxford (2013)
72. Gould, J.L., Able, K.P.: Human homing: an elusive phenomenon. Science 212(4498), 1061–1063 (1981)
73. Guilford, J.P., Zimmerman, W.S.: The Guilford-Zimmerman aptitude survey. J. Appl. Psychol. 32(1), 24–34 (1948)
74. Hafting, T., Fyhn, M., Molden, S., Moser, M.B., Moser, E.I.: Microstructure of a spatial map in the entorhinal cortex. Nature 436(7052), 801–806 (2005)
75. Hartley, T., Maguire, E.A., Spiers, H.J., Burgess, N.: The well-worn route and the path less traveled: distinct neural bases of route following and wayfinding in humans. Neuron 37(5), 877–888 (2003)
76. Hartley, T., Trinkler, I., Burgess, N.: Geometric determinants of human spatial memory. Cognition 94(1), 39–75 (2004)
77. Heft, H.: Way-finding as the perception of information over time. Popul. Environ. 6(3), 133–150 (1983)
78. Hegarty, M., Waller, D.: Individual differences in spatial abilities. In: Shah, P., Miyake, A. (eds.) The Cambridge Handbook of Visuospatial Thinking, pp. 121–169. Cambridge University Press, Cambridge (2005)
79. Hegarty, M., Richardson, A.E., Montello, D.R., Lovelace, K.L., Subbiah, I.: Development of a self-report measure of environmental spatial ability. Intelligence 30(5), 425–447 (2002)
80. Hegarty, M., Montello, D.R., Richardson, A.E., Ishikawa, T., Lovelace, K.L.: Spatial abilities at different scales: individual differences in aptitude-test performance and spatial-layout learning. Intelligence 34(2), 151–176 (2006)
81. Herskovits, A.: Language and Spatial Cognition. Cambridge University Press, Cambridge (1986)
82. Heye, C., Rüetschi, U.J., Timpf, S.: Komplexität von Routen in öffentlichen Verkehrssystemen. In: Strobl, J., Blaschke, T., Griesebner, G. (eds.) Angewandte Geographische Informationsverarbeitung XV, pp. 159–168. Wichmann, Heidelberg (2003)
83. Hirtle, S.C., Jonides, J.: Evidence of hierarchies in cognitive maps. Mem. Cognit. 13(3), 208–217 (1985)
84. Hobbs, J.R.: Granularity. In: Joshi, A.K. (ed.) Proceedings of the 9th International Joint Conference on Artificial Intelligence, pp. 432–435. Morgan Kaufmann, Los Angeles (1985)
85. Hochmair, H.: Investigating the effectiveness of the least-angle strategy for wayfinding in unknown street networks. Environ. Plann. B Plann. Des. 32(5), 673–691 (2005)
86. Hochmair, H., Frank, A.U.: Influence of estimation errors on wayfinding-decisions in unknown street networks: analyzing the least-angle strategy. Spat. Cognit. Comput. 2(4), 283–313 (2002)
87. Ishikawa, T., Montello, D.R.: Spatial knowledge acquisition from direct experience in the environment: individual differences in the development of metric knowledge and the integration of separately learned places. Cognit. Psychol. 52(2), 93–129 (2006)
88. Ishikawa, T., Nakamura, U.: Landmark selection in the environment: relationships with object characteristics and sense of direction. Spat. Cognit. Comput. 12(1), 1–22 (2012)
89. Ishikawa, T., Fujiwara, H., Imai, O., Okabe, A.: Wayfinding with a GPS-based mobile navigation system: a comparison with maps and direct experience. J. Environ. Psychol. 28, 74–82 (2008)
90. Janelle, D.G.: Impact of information technologies. In: Hanson, S., Giuliano, G. (eds.) The Geography of Urban Transportation, pp. 86–112. Guilford Press, New York (2004)
91. Janzen, G., Jansen, C., Turennout, M.v.: Memory consolidation of landmarks in good navigators. Hippocampus 18(1), 40–47 (2008)

92. Janzen, G., Turennout, M.v.: Selective neural representation of objects relevant for navigation. Nat. Neurosci. **7**(6), 673–677 (2004)
93. Johnson-Laird, P.N.: Mental Models: Towards a Cognitive Science of Language, Inference and Consciousness. Cambridge University Press, Cambridge, UK (1983)
94. Kahneman, D.: Thinking, Fast and Slow. Farrar, Straus and Giroux, New York (2011)
95. Kaplan, S., Kaplan, R.: Cognition and Environment: Functioning in an Uncertain World. Praeger, New York (1983)
96. Keeton, W.T.: Magnets interfere with pigeon homing. Proc. Natl. Acad. Sci. **68**(1), 102–106 (1971)
97. Kelly, J.W., McNamara, T.P., Bodenheimer, B., Carr, T.H., Rieser, J.J.: The shape of human navigation: how environmental geometry is used in maintenance of spatial orientation. Cognition **109**(2), 281–286 (2008)
98. Kelly, J.W., McNamara, T.P., Bodenheimer, B., Carr, T.H., Rieser, J.J.: Individual differences in using geometric and featural cues to maintain spatial orientation: cue quantity and cue ambiguity are more important than cue type. Psychonomic Bull. Rev. **16**(1), 176–181 (2009)
99. Klatzky, R.L.: Allocentric and egocentric spatial representations: definitions, distinctions, and interconnections. In: Freksa, C., Habel, C., Wender, K.F. (eds.) Spatial Cognition, Lecture Notes in Artificial Intelligence, vol. 1404, pp. 1–17. Springer, Berlin (1998)
100. Klein, W.: Wegauskünfte. Zeitschrift für Literaturwissenschaft und Linguistik **33**, 9–57 (1979)
101. Kleinfeld, J.: Visual memory in village Eskimo and urban Caucasian children. Arctic **24**(2), 132–138 (1971)
102. Klippel, A., Montello, D.R.: Linguistic and non-linguistic turn direction concepts. In: Winter, S., Duckham, M., Kulik, L., Kuipers, B. (eds.) Spatial Information Theory, Lecture Notes in Computer Science, vol. 4736, pp. 354–372. Springer, Berlin (2007)
103. Klippel, A., Tappe, H., Habel, C.: Pictorial representations of routes: chunking route segments during comprehension. In: Freksa, C., Brauer, W., Habel, C., Wender, K.F. (eds.) Spatial Cognition III, Lecture Notes in Artificial Intelligence, vol. 2685, pp. 11–33. Springer, Berlin (2003)
104. Knauff, M.: Space to Reason: A Spatial Theory of Human Thought. MIT Press, Cambridge (2013)
105. Knauff, M., Ragni, M.: Cross-cultural preferences in spatial reasoning. J. Cognit. Cult. **11**(1), 1–21 (2011)
106. Kozhevnikov, M., Hegarty, M.: A dissociation between object manipulation spatial ability and spatial orientation ability. Mem. Cognit. **29**(5), 745–756 (2001)
107. Kuhn, W.: Ontology of landscape in language. In: Mark, D.M., Turk, A.G., Burenhult, N., Stea, D. (eds.) Landscape in Language: Transdisciplinary Perspectives. Culture and Language Use, vol. 4, pp. 369–379. John Benjamins Publishing Company, Philadelphia (2011)
108. Kuhn, W., Frank, A.U.: A formalization of metaphors and image-schemas in user interfaces. In: Mark, D.M., Frank, A.U. (eds.) Cognitive and Linguistic Aspects of Geographic Space. NATO ASI Series D: Behavioural and Social Sciences, vol. 63, pp. 419–434. Kluwer Academic Publishers, Dordrecht (1991)
109. Kuipers, B.J.: Modeling spatial knowledge. Cognit. Sci. **2**(2), 129–153 (1978)
110. Kuipers, B.J.: On representing commonsense knowledge. In: Findler, N.V. (ed.) Associative Networks: Representation and Use of Knowledge by Computers, pp. 393–408. Academic, New York (1979)
111. Lakoff, G.: Women, Fire, and Dangerous Things: What Categories Reveal About the Mind. The University of Chicago Press, Chicago (1987)
112. Lakoff, G., Johnson, M.: Metaphors We Live By. The University of Chicago Press, Chicago (1980)
113. Lakoff, G., Johnson, M.: Philosophy in the Flesh: The Embodied Mind and Its Challenge to Western Thought. Basic Books, New York (1999)
114. Landau, B., Jackendoff, R.: "What" and "where" in spatial language and spatial cognition. Behav. Brain Sci. **16**(2), 217–238 (1993)

115. Larkin, J.H., Simon, H.A.: Why a diagram is (sometimes) worth ten thousand words. Cognit. Sci. 11(1), 65–100 (1987)
116. Lee, P.U., Tversky, B.: Interplay between visual and spatial: the effect of landmark descriptions on comprehension of route/survey spatial descriptions. Spat. Cognit. Comput. 5(2, 3), 163–185 (2005)
117. Lee, P.U., Tappe, H., Klippel, A.: Acquisition of landmark knowledge from static and dynamic presentation of route maps. In: 24th Annual Meeting of the Cognitive Science Society. George Mason University, Fairfax, Virginia (2002)
118. Levelt, W.J.M.: Perspective taking and ellipsis in spatial descriptions. In: Bloom, P., Peterson, M.A., Nadel, L., Garrett, M.F. (eds.) Language and Space, pp. 77–108. The MIT Press, Cambridge (1996)
119. Levine, M., Jankovic, I.N., Palij, M.: Principles of spatial problem solving. J. Exp. Psychol. Gen. 111(2), 157–175 (1982)
120. Levinson, S.C.: Space in Language and Cognition. Cambridge University Press, Cambridge (2003)
121. Lewicki, P., Hill, T., Czyzewska, M.: Nonconscious acquisition of information. Am. Psychol. 47(6), 796–801 (1992)
122. Liben, L.S., Myers, L.J.: Developmental changes in children's understanding of maps: What, when, and how? In: Plumert, J.M., Spencer, J.P. (eds.) The Emerging Spatial Mind, pp. 193–218. Oxford University Press, Oxford (2007)
123. Likert, R.: A technique for the measurement of attitudes. Archives Psychol. 140, 1–55 (1932)
124. Lloyd, R., Patton, D., Cammack, R.: Basic-level geographic categories. Prof. Geogr. 48(2), 181–194 (1996)
125. Loomis, J.M., Klatzky, R.L., Golledge, R.G., Cicinelli, J.G., Pellegrino, J.W., Fry, P.A.: Nonvisual navigation by blind and sighted: assessment of path integration ability. J. Exp. Psychol. Gen. 122(1), 73–91 (1993)
126. Loomis, J.M., Klatzky, R.L., Golledge, R.G., Philbeck, J.W.: Human navigation by path integration. In: Golledge, R.G. (ed.) Wayfinding Behavior, pp. 125–151. The John Hopkins University Press, Baltimore (1999)
127. Lovelace, K.L., Hegarty, M., Montello, D.R.: Elements of good route directions in familiar and unfamiliar environments. In: Freksa, C., Mark, D.M. (eds.) Spatial Information Theory, Lecture Notes in Computer Science, vol. 1661, pp. 65–82. Springer, Berlin (1999)
128. Lynch, K.: The Image of the City. The MIT Press, Cambridge (1960)
129. Maguire, E.A., Burgess, N., Donnett, J.G., Frackowiak, R.S.J., Frith, C.D., O'Keefe, J.: Knowing where and getting there: a human navigation network. Science 280(5365), 921–924 (1998)
130. Maguire, E.A., Gadian, D.G., Johnsrude, I.S., Good, C.D., Ashburner, J., Frackowiak, R.S.J., Frith, C.D.: Navigation-related structural change in the hippocampi of taxi drivers. Proc. Natl. Acad. Sci. 97(8), 4398–4403 (2000)
131. Mandler, J.M.: Representation. In: Damon, W. (ed.) Handbook of Child Psychology, vol. 2, pp. 255–308. Wiley, Hoboken (1998)
132. Mandler, J.M.: On the spatial foundations of the conceptual system and its enrichment. Cognit. Sci. 36(3), 421–451 (2012)
133. Mark, D.M.: Finding simple routes: "ease of description" as an objective function in automated route selection. In: Second Symposium on Artificial Intelligence Applications, pp. 577–581. IEEE, Miami Beach (1985)
134. Mark, D.M.: Toward a theoretical framework for geographic entity types. In: Frank, A.U., Campari, I. (eds.) Spatial Information Theory, Lecture Notes in Computer Science, vol. 716, pp. 270–283. Springer, Berlin (1993)
135. Mark, D.M., Egenhofer, M.J.: Calibrating the meanings of spatial predicates from natural language: Line-region relations. In: Waugh, T.C., Healey, R.G. (eds.) Advances in GIS Research: Proceedings of 6th International Symposium on Spatial Data Handling, pp. 538–553. Edinburgh (1994)

136. Mark, D.M., Turk, A.G.: Landscape categories in Yindjibarndi. In: Kuhn, W., Worboys, M.F., Timpf, S. (eds.) Spatial Information Theory, Lecture Notes in Computer Science, vol. 2825, pp. 28–45. Springer, Berlin (2003)

137. Mark, D.M., Turk, A.G., Stea, D.: Progress on Yindjibarndi ethnophysiography. In: Winter, S., Duckham, M., Kulik, L., Kuipers, B. (eds.) Spatial Information Theory, Lecture Notes in Computer Science, vol. 4736, pp. 1–19. Springer, Berlin (2007)

138. Marr, D.: Vision. W. H. Freeman and Company, New York (1982)

139. Mathis, A., Herz, A.V.M., Stemmler, M.B.: Resolution of nested neuronal representations can be exponential in the number of neurons. Phys. Rev. Lett. 109(1), 018,103 (2012)

140. McGee, M.G.: Human Spatial Abilities: Sources of Sex Differences. Praeger, New York (1979)

141. McNamara, T.P.: Mental representations of spatial relations. Cognit. Psychol. 18(1), 87–121 (1986)

142. Michon, P.E., Denis, M.: When and why are visual landmarks used in giving directions? In: Montello, D.R. (ed.) Spatial Information Theory, Lecture Notes in Computer Science, vol. 2205, pp. 292–305. Springer, Berlin (2001)

143. Milgram, S., Jodelet, D.: Psychological maps of Paris. In: Proshansky, H.M., Ittelson, W.H., Rivlin, L. (eds.) Environmental Psychology: People and Their Physical Settings, 2nd edn., pp. 104–124. Holt, Rinehart and Winston, New York (1976)

144. Miller, G.A.: The magical number seven, plus or minus two: some limits on our capacity for processing information. Psychol. Rev. 63, 81–97 (1956)

145. Mittelstaedt, M.L., Mittelstaedt, H.: Homing by path integration in a mammal. Naturwissenschaften 67(11), 566–567 (1980)

146. Montello, D.R.: Scale and multiple psychologies of space. In: Frank, A.U., Campari, I. (eds.) Spatial Information Theory, Lecture Notes in Computer Science, vol. 716, pp. 312–321. Springer, Berlin (1993)

147. Montello, D.R.: A new framework for understanding the acquistion of spatial knowledge in large-scale environments. In: Egenhofer, M.J., Golledge, R.G. (eds.) Spatial and Temporal Reasoning in Geographic Information Systems, chap. 11, pp. 143–154. Oxford University Press, New York (1998)

148. Montello, D.R.: Spatial cognition. In: Smelser, N.J., Baltes, P.B. (eds.) International Encyclopedia of the Social and Behavioral Sciences, pp. 14,771–14,775. Pergamon Press, Oxford (2001)

149. Montello, D.R.: Navigation. In: Shah, P., Miyake, A. (eds.) Handbook of Visuospatial Thinking, pp. 257–294. Cambridge University Press, Cambridge (2005)

150. Montello, D.R.: You are where? the function and frustration of you-are-here (YAH) maps. Spat. Cognit. Comput. 10(2), 94–104 (2010)

151. Montello, D.R., Raubal, M.: Functions and applications of spatial cognition. In: Waller, D., Nadel, L. (eds.) The APA Handbook of Spatial Cognition, pp. 249–264. American Psychological Association, Washington, DC (2013)

152. Müller, M., Wehner, R.: Path integration in desert ants, cataglyphis fortis. Proc. Natl. Acad. Sci. 85(14), 5287–5290 (1988)

153. Munnich, E., Landau, B., Dosher, B.A.: Spatial language and spatial representation: a cross-linguistic comparison. Cognition 81(3), 171–208 (2001)

154. Newcombe, N., Frick, A.: Early education for spatial intelligence: why, what, and how. Mind Brain Educ. 4(3), 102–111 (2010)

155. Newcombe, N., Huttenlocher, J.: Development of spatial cognition. In: Damon, W., Lerner, R.M. (eds.) Handbook of Child Psychology: Theoretical Models of Human Development, vol. 2, 6th edn., pp. 734–776. Wiley, Hoboken (2006)

156. Newcombe, N., Uttal, D.H., Sauter, M.: Spatial development. In: Zelazo, P.D. (ed.) The Oxford Handbook of Developmental Psychology, vol. 1. Oxford University Press, Oxford (2013)

157. Ogden, C.K., Richards, I.A.: The Meaning of Meaning: A Study of the Influence of Language Upon Thought and of the Science of Symbolism. Routledge & Kegan Paul, London (1923)

158. O'Keefe, J., Dostrovsky, J.: The hippocampus as a spatial map: preliminary evidence from unit activity in the freely-moving rat. Brain Res. **34**(1), 171–175 (1971)
159. O'Keefe, J., Nadel, L.: The Hippocampus as a Cognitive Map (definition of a cognitive map). Clarendon Press, Oxford (1978)
160. Olson, D.L., Bialystok, E.: Spatial Cognition. Child Psychology. Lawrence Erlbaum Associates, Hillsdale (1983)
161. O'Regan, J.K.: Why Red Doesn't Sound Like a Bell: Understanding the Feel of Consciousness. Oxford University Press, New York (2011)
162. Parush, A., Berman, D.: Navigation and orientation in 3D user interfaces: the impact of navigation aids and landmarks. Int. J. Hum. Comput. Stud. **61**(3), 375–395 (2004)
163. Pazzaglia, F., De Beni, R.: Strategies of processing spatial information in survey and landmark-centred individuals. Eur. J. Cognit. Psychol. **13**(4), 493–508 (2001)
164. Pazzaglia, F., Meneghetti, C.: Spatial text processing in relation to spatial abilities and spatial styles. J. Cognit. Psychol. **24**(8), 972–980 (2012)
165. Piaget, J.: Studies in Reflecting Abstraction. Psychology Press, London, UK (2000)
166. Piaget, J., Inhelder, B.: The Child's Conception of Space. Routledge & Kegan Paul, London (1956)
167. Pick, H., Acredolo, L. (eds.): Spatial Orientation: Theory, Research, and Application. Plenum Press, New York (1983)
168. Pinker, S.: The Language Instinct: How the Mind Creates Language, 2nd edn. Harper Perennial Modern Classics, New York (2007)
169. Plumert, J.M., Carswell, C., DeVet, K., Ihrig, D.: The content and organization of communication about object locations. J. Mem. Lang. **34**, 477–498 (1995)
170. Plumert, J.M., Spalding, T.L., Nichols-Whitehead, P.: Preferences for ascending and descending hierarchical organization in spatial communication. Mem. Cognit. **29**(2), 274–284 (2001)
171. Purves, D., Lotto, R.B.: Why We See What We Do Redux. Sinauer Associates, Inc., Sunderland (2011)
172. Raubal, M., Egenhofer, M.: Comparing the complexity of wayfinding tasks in built environments. Environ. Plann. B **25**(6), 895–913 (1998)
173. Redish, A.D.: Beyond the Cognitive Map: From Place Cells to Episodic Memory. The MIT Press, Cambridge (1999)
174. Redish, A.D., Ekstrom, A.D.: Hippocampus and related areas: what the place cell literature tells us about cognitive maps in rats and humans. In: Waller, D., Nadel, L. (eds.) Handbook of Spatial Cognition, pp. 15–34. American Psychological Association, Washington, DC (2013)
175. Relph, E.C.: Place and Placelessness. Pion Ltd., London (1976)
176. Richter, K.F., Winter, S.: Harvesting user-generated content for semantic spatial information: the case of landmarks in OpenStreetMap. In: Hock, B. (ed.) Proceedings of the Surveying and Spatial Sciences Biennial Conference 2011, pp. 75–86. Surveying and Spatial Sciences Institute, Wellington (2011)
177. Richter, D., Richter, K.F., Winter, S.: The impact of classification approaches on the detection of hierarchies in place descriptions. In: Vandenbroucke, D., Bucher, B., Crompvoets, J. (eds.) Geographic Information Science at the Heart of Europe, Lecture Notes in Geoinformation and Cartography, pp. 191–206. Springer, Berlin (2013)
178. Richter, D., Vasardani, M., Stirling, L., Richter, K.F., Winter, S.: Zooming in–zooming out: hierarchies in place descriptions. In: Krisp, J.M. (ed.) Progress in Location-Based Services, Lecture Notes in Geoinformation and Cartography. Springer, Berlin (2013)
179. Rieser, J.J.: Spatial orientation of six-month-old infants. Child Dev. **50**(4), 1078–1087 (1979)
180. Rosch, E., Mervis, C.B., Gray, W.D., Johnson, D.M., Boyes-Braem, P.: Basic objects in natural categories. Cognit. Psychol. **8**(3), 382–439 (1976)
181. Sacks, O.: The Man Who Mistook His Wife for a Hat. Picador, London (1985)
182. Sadalla, E.K., Magel, S.: The perception of traversed distance. Environ. Behav. **12**(1), 65–79 (1980)
183. Sadalla, E.K., Staplin, L.J.: An information storage model for distance cognition. Environ. Behav. **12**(2), 183–193 (1980)

184. Sadalla, E.K., Burroughs, J., Staplin, L.J.: Reference points in spatial cognition. J. Exp. Psychol. Hum. Learn. Mem. **6**(5), 516–528 (1980)
185. Samet, H.: The Design and Analysis of Spatial Data Structures. Addison-Wesley, Reading (1990)
186. Sanchez, C.A., Branaghan, R.J.: The interaction of map resolution and spatial abilities on map learning. Int. J. Hum. Comput. Stud. **67**(5), 475–481 (2009)
187. Schegloff, E.A.: Notes on a conversational practice: formulating place. In: Sudnow, D. (ed.) Studies in Social Interaction, vol. 75, pp. 75–119. MacMillan, New York (1972)
188. Schelling, T.C.: The Strategy of Conflict. Harvard University Press, Cambridge (1960)
189. Schmid, F., Richter, K.F., Peters, D.: Route aware maps: multigranular wayfinding assistance. Spat. Cognit. Comput. **10**(2), 184–206 (2010)
190. Schmid, F., Kuntzsch, C., Winter, S., Kazerani, A., Preisig, B.: Situated local and global orientation in mobile you-are-here maps. In: de Sa, M., Carrico, L., Correia, N. (eds.) 12th International Conference on Human Computer Interaction with Mobile Devices and Services (MobileHCI), pp. 83–92. ACM Press, Lisbon (2010)
191. Schneider, G.E.: Two visual systems. Science **163**(3870), 895–902 (1969)
192. Searle, J.R.: Speech Acts. Cambridge University Press, Cambridge (1969)
193. Searle, J.R.: Minds, brains, and programs. Behav. Brain Sci. **3**, 417–424 (1980)
194. Shanon, B.: Where questions. In: 17th Annual Meeting of the Association for Computational Linguistics. ACL, University of California at San Diego, La Jolla (1979)
195. Shannon, C.E., Weaver, W.: The Mathematical Theory of Communication. University of Illinois Press, Chicago (1949)
196. Shelton, A.L., Gabrieli, J.D.E.: Neural correlates of encoding space from route and survey perspectives. J. Neurosci. **22**(7), 2711–2717 (2002)
197. Shepard, R.N., Metzler, J.: Mental rotation of three-dimensional objects. Science **171**(3972), 701–703 (1971)
198. Siegel, A.W., White, S.H.: The development of spatial representations of large-scale environments. In: Reese, H. (ed.) Advances in Child Development and Behaviour, pp. 9–55. Academic, New York (1975)
199. Silverman, I., Eals, M.: Sex differences in spatial abilities: evolutionary theory and data. In: Barkow, J.H., Cosmides, L., Tooby, J. (eds.) The Adapted Mind: Evolutionary Psychology and the Generation of Culture, pp. 533–549. Oxford University Press, New York (1992)
200. Smith, B. (ed.): Foundations of Gestalt Theory. Philosphia Resources Library. Philosophia Verlag, Munich (1988)
201. Sorrows, M.E., Hirtle, S.C.: The nature of landmarks for real and electronic spaces. In: Freksa, C., Mark, D.M. (eds.) Spatial Information Theory, Lecture Notes in Computer Science, vol. 1661, pp. 37–50. Springer, Berlin (1999)
202. Sperber, D., Wilson, D.: Relevance: Communication and Cognition. Basil Blackwell, Oxford (1986)
203. Spiers, H.J., Maguire, E.A.: A navigational guidance system in the human brain. Hippocampus **17**(8), 618–626 (2007)
204. Steck, S.D., Mallot, H.A.: The role of global and local landmarks in virtual environment navigation. Presence **9**(1), 69–83 (2000)
205. Stevens, Q.: The shape of urban experience: a reevaluation of Lynch's five elements. Environ. Plann. B Plann. Des. **33**(6), 803–823 (2006)
206. Stevens, A., Coupe, P.: Distortions in judged spatial relations. Cognit. Psychol. **10**(4), 422–437 (1978)
207. Streeter, L.A., Vitello, D., Wonsiewicz, S.A.: How to tell people where to go: comparing navigational aids. Int. J. Man Mach. Stud. **22**(5), 549–562 (1985)
208. Talmy, L.: How language structures space. In: Pick, H. (ed.) Spatial Orientation Theory: Research and Application, pp. 225–282. Plenum Press, New York (1983)
209. Taube, J.S., Muller, R.U., Ranck, J.B.: Head-direction cells recorded from the postsubiculum in freely moving rats. I. Description and quantitative analysis. J. Neurosci. **10**(2), 420–435 (1990)

210. Taube, J.S., Muller, R.U., Ranck, J.B.: Head-direction cells recorded from the postsubiculum in freely moving rats. II. Effects of environmental manipulations. J. Neurosci. 10(2), 436–447 (1990)
211. Taylor, H.A., Tversky, B.: Descriptions and depictions of environments. Mem. Cognit. 20(5), 483–496 (1992)
212. Taylor, H.A., Tversky, B.: Spatial mental models derived from survey and route descriptions. J. Mem. Lang. 31(2), 261–292 (1992)
213. Tenbrink, T., Winter, S.: Variable granularity in route directions. Spat. Cognit. Comput. 9(1), 64–93 (2009)
214. Thorndyke, P.W.: Distance estimation from cognitive maps. Cognit. Psychol. 13(4), 526–550 (1981)
215. Timpf, S., Frank, A.U.: Using hierarchical spatial data structures for hierarchical spatial reasoning. In: Hirtle, S.C., Frank, A.U. (eds.) Spatial Information Theory, Lecture Notes in Computer Science, vol. 1329, pp. 69–83. Springer, Berlin (1997)
216. Tolman, E.C.: Cognitive maps in rats and men. Psychol. Rev. 55(4), 189–208 (1948)
217. Tomko, M., Winter, S.: Describing the functional spatial structure of urban environments. Comput. Environ. Urban Syst. 41, 177–187 (2013)
218. Trowbridge, C.C.: On fundamental methods of orientation and "imaginary maps". Science 38(990), 888–897 (1913)
219. Tse, D., Langston, R.F., Kakeyama, M., Bethus, I., Spooner, P.A., Wood, E.R., Witter, M.P., Morris, R.G.M.: Schemas and memory consolidation. Science 316(5821), 76–82 (2007)
220. Tversky, A., Kahneman, D.: Judgement under uncertainty: heuristics and biases. Science 185(4157), 1124–1131 (1974)
221. Tversky, B.: Cognitive maps, cognitive collages, and spatial mental models. In: Frank, A.U., Campari, I. (eds.) Spatial Information Theory, Lecture Notes in Computer Science, vol. 716, pp. 14–24. Springer, Heidelberg (1993)
222. Tversky, B., Hard, B.M.: Embodied and disembodied cognition: spatial perspective-taking. Cognition 110(1), 124–129 (2009)
223. Tversky, B., Lee, P.U.: How space structures language. In: Freksa, C., Habel, C., Wender, K.F. (eds.) Spatial Cognition, Lecture Notes in Artificial Intelligence, vol. 1404, pp. 157–175. Springer, Berlin (1998)
224. Twaroch, F.: Sandbox geography. Ph.d. thesis, Technical University Vienna (2007)
225. Vandenberg, S.G., Kuse, A.R.: Mental rotations, a group test of three-dimensional spatial visualization. Percept. Mot. Skills 47(2), 599–604 (1978)
226. Vanetti, E.J., Allen, G.L.: Communicating environmental knowledge: the impact of verbal and spatial abilities on the production and comprehension of route directions. Environ. Behav. 20(6), 667–682 (1988)
227. Varela, F.J., Thompson, E., Rosch, E.: The Embodied Mind: Cognitive Science and the Human Experience. The MIT Press, Cambridge (1991)
228. Vasardani, M., Timpf, S., Winter, S., Tomko, M.: From descriptions to depictions: a conceptual framework. In: Tenbrink, T., Stell, J., Galton, A., Wood, Z. (eds.) Spatial Information Theory, Lecture Notes in Computer Science, vol. 8116, pp. 299–319. Springer, Cham (2013)
229. Vygotsky, L.S.: Thought and Language. MIT Press, Cambridge (1986)
230. Walker, M.M., Dennis, T.E., Kirschvink, J.L.: The magnetic sense and its use in long-distance navigation by animals. Curr. Opin. Neurobiol. 12(6), 735–744 (2002)
231. Waller, D., Nadel, L. (eds.): Handbook of Spatial Cognition. American Psychological Association, Washington, DC (2013)
232. Wang, J., Schwering, A.: The accuracy of sketched spatial relations: how cognitive errors influence sketch representation. In: Tenbrink, T., Winter, S. (eds.) Proceedings of the International Workshop Presenting Spatial Information: Granularity, Relevance, and Integration, pp. 40–47. SFB/TR8 and University of Melbourne, Melbourne, Australia (2009)
233. Wang, R.F., Spelke, E.S.: Updating egocentric representations in human navigation. Cognition 77(3), 215–250 (2000)

234. Wehner, R.: Desert ant navigation: how miniature brains solve complex tasks. J. Comp. Physiol. **189**(8), 579–588 (2003)
235. Weissensteiner, E., Winter, S.: Landmarks in the communication of route instructions. In: Egenhofer, M., Freksa, C., Miller, H.J. (eds.) Geographic Information Science, Lecture Notes in Computer Science, vol. 3234, pp. 313–326. Springer, Berlin (2004)
236. Wertheimer, M.: Über Gestalttheorie. Philosophische Zeitschrift für Forschung und Aussprache **1**, 39–60 (1925)
237. Westheimer, G.: Gestalt theory reconfigured: Max Wertheimer's anticipation of recent developments in visual neuroscience. Perception **28**(1), 5–15 (1999)
238. Winter, S., Freksa, C.: Approaching the notion of place by contrast. J. Spat. Inf. Sci. **2012**(5), 31–50 (2012)
239. Wittgenstein, L.: Philosophical Investigations, 2nd edn. Basil Blackwell, Oxford (1963)
240. Wolbers, T., Hegarty, M., Büchel, C., Loomis, J.M.: Spatial updating: how the brain keeps track of changing object locations during observer motion. Nat. Neurosci. **11**(10), 1223–1230 (2008)
241. Wunderlich, D., Reinelt, R.: How to get there from here. In: Jarvella, R.J., Klein, W. (eds.) Speech, Place, and Action, pp. 183–201. Wiley, Chichester (1982)
242. van der Zee, E., Slack, J. (eds.): Representing Direction in Language and Space. Oxford University Press, Oxford (2003)

Chapter 4
Conceptual Aspects: How Landmarks Can Be Described in Data Models

Abstract Landmarks seem to be cross with current spatial data models. We have argued that landmarks are mental concepts having a fundamental role in forming the spatial reference frame for mental spatial representations. But landmarks are not a fundamental category in current geographic information modelling. For example, among Kuhn's list of *core concepts of spatial information* [36] one finds *location* and *objects* as separate concepts, which appears to be incompatible with our cognitively motivated view of landmarks as concepts that are providing just that: the link between recognizable objects and location anchoring. This chapter sets out to fill this gap. In order to bridge between the cognitive concept and a formal, machine readable description of the semantics of landmarks we resort to ontologies. In this formal conceptualization landmarks will be specified intentionally, as a function, or role, of entities representing geographic objects, an approach fully aligned with our intentional definition in Sect. 1.1. The intentional specification will cater for a quantitative *landmarkness*, which is also compatible with the graded notion of categories. Finally, landmarkness will be modelled with dynamic variability to cater for context.

4.1 The Purpose of Modelling Landmarks

It is time to consider the mapping of the cognitive concept of landmarks to the realm of an intelligent machine. An intelligent machine is one supporting an intelligent dialog with people about locations, directions and routes. So what does a machine need to know about the mental spatial representations of their human communication partners to hold a 'spatially intelligent' conversation? In order to address this question let us approach the challenge by the principle of exclusion.

First, a spatially intelligent machine does not need to know all the landmarks contained in the human communication partner's mental spatial representation. No human communication partner does. And this comes as a relief, since we have already made a point about the impossibility to externalize the complete mental spatial representation of an individual. Even in Turing's original definition of the intelligent machine it was the machine's behavior rather than the machine's (always finite) knowledge base that enabled the machine to pass the Turing test.

Secondly, even if completeness is not needed, an extensional approach of storing (finite) lists of objects that may serve as landmarks is insufficient. This must be the only valid conclusion from our insights on context-dependency of landmarks.

Thirdly, spatial intelligence cannot be based on what current spatial databases 'know' about the world already. Spatial databases contain vast numbers of entities over a range of types, and are certainly not limited to store further entities. This means, features representing the objects that are used as landmarks in mental spatial representations are either already in spatial databases or can be added easily. Actually, from an ontological perspective, the superset of all features in spatial databases is the universe of discourse for identifying features of landmarkness. These features represent the objects human conceptualization has found worth recognizing, for whatever reason. The machine also anchors each of these features to their location in its own spatial reference frame, such as the World Geodetic System[1] [37]. Furthermore, it is in the nature of spatial databases that these entities are collections of *shared* concepts. (Some) people have agreed on the taxonomy as well as the individual entities. The entities have frequently names allocated by some authority, such as geographic names, addresses, or institutional names. But the linking of these features to the landmarkness of the represented objects is missing. Even the current spatial databases' taxonomies do not contain a geographic concept *landmark*. Why is that? After all what we have heard in the previous chapters it has become clear that landmarks are not a geographic category comparable to *building* or *river*—notwithstanding that even these 'tangible' geographic types are hard to specify formally in their semantics due to conceptual vagueness and cultural dependency. Instead landmarks do not appear to be a geographic concept next to others. We have learned that landmarks are rather a cognitive concept, an internalized embodied experience statically linked to a location. Thus, a landmark is a role that an object—any object in geographic space, from geographic scale to table-top scale—can take. Practically this makes a landmark not a (class) type in a spatial database, but a property of the entities in a spatial database.

Before defining an agenda, it may be useful to consider whether some functions of an intelligent system can be realized already, and what the limitations are.

- For *understanding* a human-generated message current state of knowledge can parse the message, the locative expressions (references to objects) can be isolated, and location references both by type ("the library") or by instance name ("State Library") can be resolved by matching with database content [44]. There are limits, though. A reference to an object type will match with a

---

[1]https://en.wikipedia.org/wiki/World_Geodetic_System, last visited 3/1/2014.

larger number of database instances, and this ambiguity needs to be resolved. A reference to an instance by name may be ambiguous as well (how many instances of "State Library" are out there?), but also many names are not gazetteered, or stored in spatial databases. Addressing ambiguity requires two skills that are both not well developed in spatial databases, context-awareness and qualitative spatial reasoning. Context-awareness involves reasoning of the kind that a person in Melbourne, mentioning "State Library", most likely refers to the State Library of Victoria, Australia. Qualitative spatial reasoning, in contrast, involves mechanisms to interpret the qualitative spatial relationships used in the human-generated message to resolve ambiguity, such as in "The café opposite the State Library".

However, where needed for disambiguation, people also refer to properties of objects. Landmark references such as "the yellow building" or "the white-steepled church" pose additional challenges to spatial databases that typically do not capture a broad range of perceptual properties with the stored features.

- For *generating* human-like messages the system is still completely handicapped as it does not have a notion of landmarkness. It may, however, maintain extensional lists of points of interest or similar surrogates for landmarks, in which case it is capable of generating messages that look like spatially intelligent messages. However, the elements may miss the discriminatory power, the identifiability, or the relevance in a particular context we would now expect from an intelligent system.

Thus it is essential to introduce a notion of landmarkness for simulating human communication behavior. This notion will be made context-aware. Yet the issue of capturing context and modelling context awareness is not well understood and needs further research [9]. Furthermore, abilities of qualitative spatial reasoning are essential for an intelligent system. But this is another active area of research [4, 7, 47, 62].

In summary, we postulate that the notion of landmarkness should cover *types* as well as individual *entities* in spatial databases, and should characterize a collective expectation for embodied or mediated experiences of the environment. This addition to spatial databases must allow an internal reasoning and external communication behavior simulating some of the properties of mental spatial representations and abilities in working memory, such as choosing relevant reference objects, constructing hierarchical localization (e.g., [57]), switching between survey and route perspective, and appropriate qualification of quantified delimiters. A machine with these capabilities will not only be able to interpret human-generated spatial expressions but also will have the capacity to produce human-like spatial expressions. Further conversational capabilities in case of insufficient information are advantageous [48] and appear in regard to intelligent behavior even essential.

4.2 Towards a Landmark Model

In order to capture the semantics of landmarks in a formal, machine readable format we take an ontological approach [21]. An ontology is "a formal, explicit specification of a shared conceptualization" ([54], p. 184). This definition only slightly extends Gruber's most frequently cited one by adding 'explicit' [20]. Traditional ontological engineering chooses a first order language to produce explicit specifications (e.g., [16, 19]). Instead we choose a second order language. Second order languages were propagated by Frank and Kuhn for specifying formal semantics of geographic concepts (e.g., [11, 12, 14, 15]) roughly in parallel to the emergence of ontological engineering in computer science.

A second order language permits an algebraic approach to specifying a formal model of landmarks. The presented algebra, however, will stay incomplete. The purpose of this advance lies solely in a least ambiguous language to share formal concepts. The algebra stays incomplete by skipping some details (in particular, as mentioned, on modelling context), but also as it does not link to other, especially upper level ontologies. In this regard the hypernyms listed in Sect. 1.1 would have potential for further development [16]. Thus, the formal code can also be understood as groundwork for future completion. In principle such a specification is executable and can be tested for consistency.

4.2.1 Properties

Before the model (a formal, explicit specification in a machine readable format) can be presented and discussed let us consider the properties of objects[2] a landmark model must be able to reflect.

If landmarks are not a type, but a property, then any entity in the spatial database (any instance of any type) will have to have this property to some degree, or to some level of agreement. Considerations for this *landmarkness* are:

- There will be objects represented in the database that everybody will experience as so outstanding in the environment that this experience is linked with the location and stored in mental spatial representations.
- There will also be objects represented in the database that have meaning only for some people. They have landmarkness mostly for semantic reasons, such as *my home*. The spatial database may or may not know about the semantics, just like other people may know or not know where I live or work. In the mobile

---

[2]Smith encourages ontology engineers to give up the fuzzy term *concept* (or *conceptualization*). Instead, "ontologies [...] should be understood as having as their subject matter not concepts, but rather the universals and particulars which exist in reality and are captured in scientific laws" ([52], p. 73).

location-based services literature *personalization* has been discussed. Services learn from tracking and observing the behavior of a particular user with the goal to providing tailored individual information rather than staying a neutral service that provides the same information for everyone. Personalization is not limited to mobile location-based services, of course. Search engines access a user's search history in tailoring the response to their search request, which means that different people do already get different responses on the same requests [67]. Location-based services take a step further. Positioning technology on board of smartphones allows tracking of visited places, and methods of data mining and knowledge extraction from trajectories are now well known [70].

• There will be objects in the database that stand out in some contexts, but not in others. Thus, whatever *landmarkness* means it has to be considered together with context parameters. A visually outstanding object in an environment may not be known to a visually impaired person. The name or location of a bus stop may not be known to a car driver. A café in a mall may be a good anchor for indoor descriptions, but may be unsuited for car driving instructions, despite its address. A phone booth may be a suited reference to mark a turn location for a pedestrian, but may be too small in spatial granularity to safely guide a car driver. Context matters, and must be part of a landmark model from the start.

• For different human communication partners different symbols or names when referencing to landmarks may be appropriate. The first-time tourist may find a reference to a type ("at the church") more appropriate than the local, who might prefer the name ("at St Francis"). Thus, context determines not only the choice of the landmark, but also the way how to refer to it.

• As a consequence, a spatially intelligent system must be a context-aware system. It does matter whether the conversation partner is driving a car, visually impaired, sitting in a wheelchair, riding a bike, a public transport user, familiar with the environment or not, and so on. In each context some references to landmarks are more appropriate than other references. Going back to Janelle's [28] categorization of communication constraints (Table 3.1), a system architecture typically assumes a communication situation as synchronous and non-co-located. It may be aware of the human communication partner's location (as location-based services are) but it has no further sensors to explore and adapt to the communication situation. Since making a system context-aware is such a hard task (compared to ease with which people attend to capturing and considering context) the usual loophole of system designers is to devise a specialized system for each context. We call this solution the antonym of a spatially intelligent service.

With *landmarkness* being modelled as a context-dependent property a system will:

• Learn about *landmarkness* in particular contexts, and store it.
• Produce views, which are triggered by certain thresholds of *landmarkness*, generating instances of a type *landmark* in a particular context.

4.2.2 An Algebraic Landmark Model

We choose an algebraic approach for specifying a formal model of landmarks. An algebra consists of a type, a set of operations declared on the individuals of this domain, and axioms specifying the semantics of these operations.

> A simple example for an algebra is the algebra of *natural numbers*, let us say including 0. The operation of addition of two individuals *a* and *b*, (+) a b, is fully specified by the following axioms:
>   neutral element: (+) a 0 = a
>   associativity: (+) ((+) a b) c = (+) a ((+) b c)
>   commutativity: (+) a b = (+) b a
>   increment: i (0) =1 and (+) a i (b) = i ((+) a b).

In our context an algebraic specification will cater for the type *landmark*, derived from a property of landmarkness, and we will sketch and discuss operations on landmarkness and their behavior. Perhaps we should mention here that this formal model is independent from the language of the conversation, i.e., graphical or verbal, in the same way as the algebra of natural numbers is independent from a notation in arabic or roman numerals.

The formal model will be written here in the syntax of the functional programming language *Haskell*.[3] We do not explain the details of the language here, but even the unfamiliar reader should get the idea of the model from this approach. Haskell provides an elegant way of formal modelling. It is fully typed but supports polymorphism, e.g., talking about landmarkness of any type of entity in the spatial database. Since it does not only specify operations (interfaces) but also their semantics, Haskell, or more generally, functional programming languages are ideal tools for specification and rapid prototyping, and have been successfully applied in the geographic information domain for a while [5, 13–15, 58, 65].

For a start let us collect some data types that capture properties we have discussed before. Landmarkness is a property of any entity to a degree, thus we first define a data type representing landmarkness that we will attach later to all entities:

```
type Lns = Double
```
– a context-specific fuzzy membership value of landmarkness ([0 . . . 1])

---

[3]http://www.haskell.org/, last visited 3/1/2014.

So far, the data type is only a synonym for a double precision floating point number. This means it inherits already some semantics from the type system of Haskell. For example, addition, multiplication and subtraction are defined. This is clearly more than we want as we have not indicated any use for these operations in all discussion so far. But we might want to have some of its properties, for example its quantitativeness, compared to a Boolean variable, to express a degree of landmarkness, and the order relation to select the entity with stronger landmarkness.

As the Haskell comment, rather than the type, indicates we assume values are limited between $0 \leq \texttt{Lns} \leq 1$. This assumption needs to be formally specified elsewhere which we skip here. But let us discuss what this voluntary limitation means:

- Entities representing objects that have no landmarkness, have a property of type `Lns` of value 0, rather than no property of type `Lns`. This means, semantically `Lns` is a fuzzy membership value[4] [69]. The semantics of a fuzzy membership value is a degree of membership to a vaguely defined category. Accordingly a `Lns` property of value 0 expresses no degree of membership at all to the category *landmark*.
- If `Lns` is a fuzzy membership value to the category *landmark*, its maximum is 1, indicating a central element, or prototype, of the category *landmark*. Generally, a larger value indicates a more central element in the category.

Especially the first condition helps distinguishing between objects of no landmarkness and objects of unknown landmarkness. Objects of unknown landmarkness have no landmarkness attribute in the spatial database (an empty list).

We will specify fuzzy membership functions—functions assigning fuzzy membership values to entities [69]—in the following chapter. However, from our discussion of spatial cognition we have already learned that there is no global measure of landmarkness. The landmarkness of an object is context-dependent, and context is an open complex system including the person, the situation, and the task at hand. Hence, a fuzzy membership function, as well as a data model storing *landmarkness*, requires integration with another property, the context under which an object has landmarkness. As two very prominent ambassadors of embodied cognition, the two neurobiologists Maturana and Varela have both confirmed the view expressed in the previous chapter that cognitive representations are not objective

---

[4]Can a landmarkness value be alternatively interpreted in a probabilistic manner? Probabilities cover the same range of values, but have a significantly different meaning. From a probabilistic perspective, one could argue, the value of type Lns represents the frequency with which a reference to this object is chosen in verbal route descriptions. After the discussion above, we should add that it is chosen in verbal route descriptions *in a given communication context*. For example, Denis' skeletal descriptions came out as the descriptions containing the references everybody used in a given context. References in these descriptions would be *most likely* landmarks (as well as by broadest agreement, i.e., in the fuzzy membership sense). Hence, probabilistic interpretations are possible, and some of the computation methods in the next chapter may actually apply a probabilistic interpretation.

representations of an a priori given external world. Varela wrote: "Precisely the greatest ability of all living cognition is, within broad limits, to *pose* the relevant issues to be addressed at each moment of our life. They are not pre-given, but *enacted* or *brought forth* from a background, and what counts as relevant is what our common sense sanctions as such, always in a contextual way [...]. If the world we live in is brought forth rather than pre-given, the notion of representation cannot have a central role any longer" ([59], p. 250f.). Similarly, Maturana said earlier [38] that already the idea of a world *out there* implies a realm that preexists its construction by an observer. Thus the reality out there comes into existence for a living being only through (embodied) interaction.

Just as an example let us return to Sadalla et al. [50], who reminded that landmarks have multiple roles and are used in different spatial tasks: "The term has been used to denote (a) discriminable objects of a route, which signal navigational decisions; (b) discriminable objects of a region, which allow a subject to maintain a general geographic orientation; and (c) salient information in a memory task. These different referents suggest that landmarks may play a role in a variety of spatial abilities" (p. 516). Thus, the Eiffel Tower, which had been previously called generously a global landmark, may have low landmarkness according to (a), because of its location and its quite extended footprint, high landmarkness according to (b), because it is highly visible from many locations in Paris, and also high landmarkness according to (c) because ... well, just everybody remembers or seems to know the Eiffel Tower.

This brings up a dilemma. There is no (complete) formal model of landmarkness without context, but context, as an open-ended and arbitrary-dimensional system cannot be captured completely in a formal model. Even worse, if context cannot be captured completely then differences between computed or assigned landmarkness values, especially marginal ones, may be misleading or meaningless. So after all, what *can* we model and state about landmarkness?

With hard to defend differences in the quantity of landmarkness, the fundamental basis of above's landmarkness value remains the distinction between (a) objects with landmarkness (in some context), (b) objects with no landmarkness (in this context), and (c) objects with unknown landmarkness (in this context). The rest is a matter of grading. And grading is, considering the aim of this landmark model, actually irrelevant. The purpose of the model is two-fold: supporting the machine interpretation of human place or route descriptions, and supporting the machine generation of human-like place or route descriptions. If I am asking passers-by in a street for a route I will get a different description from each one of them. Each communication context is different because in each communication I speak to a different person. They may describe the same route (perhaps the only possible or reasonable one in this environment), but they may refer to different landmarks, and to different numbers of landmarks. The speakers choose from their mental spatial representation what they find relevant in the given situation. Furthermore, by taking into consideration their description in my wayfinding process my attention would be already primed to search for the objects referred to. I will not look out for alternative, possibly more outstanding objects and wonder why they have not been used, which would be as much a wasted effort as it is for the speaker to consciously weigh between all available landmarks in their mind. In this situation there is no *optimal* description. Many of them (ideally all of them) would help me finding the way, and thus, pragmatically they would have the same information content. Some of them may even have the same smallest number of references, satisfying brevity. Still their references may be different. One may relate a turn to the location of the Post Office, while the other prefers the fast food restaurant of global brand at the same intersection. If both would be successful for the wayfinder, then clearly none is better than the other. Numerical differences in landmarkness do not matter as long as the task can be successfully completed. Choosing landmarks is a matter of sufficing rather than optimizing; and since this is the case for a human speaker, it can be the case for the intelligent machine as well. In summary, we will complete our formal model by making landmarkness context-dependent, and by acknowledging that an object of some landmarkness in a given context is sufficient for communication in this context.

Modelling context in depth goes far beyond the scope of this book, and modelling context comprehensively is even impossible. Context could be modelled in a hierarchic taxonomy, for example, starting at top level with the distinction of spatial tasks made by Sadalla et al. : (a) landmarks in routing tasks, (b) landmarks in orienting tasks, and (c) landmarks in memorizing tasks, e.g., for places or events. Each of these categories could further be split, for example, landmarks in routing tasks will be different for different modes of travelling (because of different embodied experiences of the environment). Further levels of context specifiers can be introduced: the particular individual with their personal mental spatial representation, or the set of people this individual could be assigned to by their shared degree of familiarity with the environment (e.g., first time visitor versus local expert), or the language and cultural background of the communication partner (e.g., familiarity with medieval city outlays versus familiarity with new world grid outlays), or the time of the day (e.g., daylight versus lights at nighttime). As mentioned, the list (or hierarchy) is open-ended.

Just to illustrate how landmarkness can be modelled in a context-dependent way, let us set up a simplistic model that may even violate hierarchical groupings deliberately:

```
data Cid = Pedestrian | Motorist | Tom
  – context identifier for a number of predefined contexts
```

In Haskell, the vertical bar means an exclusive disjunction. The context identifier can take one of these alternative values, and thus, can describe landmarkness of an object either for a pedestrian, or for a motorist, or for Tom, an individual. This simplistic model will allow at least a context-dependent storage and reasoning about landmarkness. A smarter machine would replace this model by a model with more elements or more (hierarchical) structure.

With the help of a third data type the model will be enabled to refer to an object in a flexible, context-dependent manner:

```
type Ref = [Char]
   – a context-specific reference
```

With these three types we can introduce a complex data type describing the context-dependent landmarkness property of an entity in a spatial database:

```
data Ls = Ls cid :: Cid, lns :: Lns, ref :: Ref
   – a context-dependent landmarkness property: a triplet
   – any entity can have multiple of these properties
```

Note that any entity in a spatial database can have multiple of these properties, but only one for each *Cid*. This means entities have lists of *Ls* with *Cid* as unique key:

```
data Entity = Entity {(anything), lnss :: [Ls]}
```

Entities can have any structure. Typically it will be a type from some database taxonomy, an identifier, an official name, a geometric description as a point, polyline, polygon, volume or a set of these, and further thematic descriptors. For readability we have generalized this structure in the code and concentrated on the relevant bit. All what we did was adding one property, a list of landmarknesses in different contexts. Here is an example of what we can represent now:

```
Entity ... [Ls Pedestrian 0.7 "St Francis",
   Ls Motorist 0.9 "the church"]
```

This database entity is a good landmark for motorists, for whom "the church" should be an appropriate reference in verbal communication. It will also be a good landmark, despite its spatial extension, for pedestrians, for whom it is easy to identify the building in situ as "St Francis". The database has not yet stored any information about the suitedness to refer to this building in communication with Tom personally, but of course Tom can be identified as a pedestrian. Also the model has not yet been equipped with an ability to store information about the suitedness of this reference for users of public transport. This would require an extension of *Cid* by another context identifier.

In a similar vein we add to the abstract class of *Entities* to specify operations on these properties:

```
class Entities e where
    addLns :: e -> Ls -> e
    getLnss :: e -> [Ls]
    getLns :: e -> Cid -> [Ls]
    hasLns :: e -> Cid -> Bool
```

Among these operations, *addLns* should allow to add a landmarkness property to an entity of any type (in Haskell, *e* is a type variable). Accordingly, *getLns* should provide access to the landmarkness for the context *Cid*, or alternatively, *getLnss* should provide access to the full list of stored landmarkness properties of an individual database entity.

For the particular data type *Entity* the semantics of these operations can be specified as:

```
instance Entities Entity where
    addLns (Entity ... l) ls = Entity ... (l++[ls])
    getLnss e = lnss e
    getLns e c = [x | x <- (lnss e), (cid x==c)]
    hasLns e c = if (getLns e c == [])
        then False
        else if (lval (head (getLns e c)) > 0)
            then True
            else False
```

The first operation copies all content from the entity and appends the new landmarkness at the end of the existing list of landmarknesses. This operation works even if *l* was prior an empty list. The second operation is only an alias for the observer that had been introduced with the data type *Entity*. The third one goes through all elements of the list of landmarknesses of this entity, and when it finds an element for the context *Cid* it puts it into the output list. Note that *Cid* should be a unique key in an entity's list, such that the result is either a list of one element or an empty list. The fourth operation, *hasLns*, takes the output of *getLns*—i.e., either an empty list or a list of one element—and tests whether there is a context-related landmarkness. Only if *lval*> 0 it returns *True*, and in both alternative cases it returns *False*. Either there is no landmarkness, *lval*= 0, or the landmarkness is unknown and getLns returns an empty list.

The property of landmarkness can also be mapped on a list of landmarks, by calling all entities that have some landmarkness under a given context a *landmark*:

```
class Landmarks e where
    collect :: [e] -> Cid -> [e]

instance Landmarks Entity where
    collect el c =
        [ e | e <- el, hasLns e c ]
```

Also, with access to the geometric description of entities in the spatial database, an operation can be defined to select all context-related landmarks within an area of interest, for example within a buffer zone around a route.

With some more effort such an operator can even be extended for lifting the landmarkness property from entities to types. For example, if all entities of a type turn out to have some landmarkness then the type could be said to have landmarkness. Accordingly, one of the fuzzy membership functions we will introduce in the next section is based on type landmarkness, generalizing instance properties of landmarkness.

Admittedly some properties of landmarks that we had discussed in the previous chapter are not yet captured in this formal model: hierarchies of salience, hierarchies of spatial granularities, and qualitative spatial relations in some flexible spatial frames of reference. The reason why these (necessary) properties are not put into the model here is that they are expected to appear from the general structure of a spatial database in their current forms. Entities in spatial databases have a type, a thematic description, and a geometric description in a spatial frame of reference [40]. The database itself comes with a taxonomy. Especially for integration of different databases, where taxonomies typically clash, taxonomies can be created in an ad-hoc manner, for example based on similarity computed from affordance [29, 30]. Taxonomies of entity types provide a specialization hierarchy, or an *is-a hierarchy*. Specialization, especially when described by affordance, will be a key of context-dependent choice of landmarks, in addition to our formal model's capacity to assign different landmarkness measures to each entity depending on context. Some methods of landmark identification are solely type-based (see Chap. 5). Hierarchies of spatial granularity (*part-of* or containment hierarchies) are expected to result from the geometric description of the entities. And qualitative spatial relations between landmarks, or relata, are intertwined with the selection process of landmarks (Chap. 6), which use either the geometric descriptions of entities again, or use one of the qualitative spatial calculi [62]. Thus, the presented model should be sufficient so far.

Chapter 5 will be dedicated to compute the values of landmarkness in this model. In principle multiple approaches are possible: studying the properties of objects that can be sensed or learned, category-based classifications, studying the graphic or verbal expressions of people searching for relata, or direct surveying via questionnaires or user-contributed content. Each of these methods applies to a particular context—e.g., a survey among pedestrians will reveal another list as a survey among motorists, and the context can now be stored and accessed together with the landmarkness value. This means if the goal is collecting the landmarks of an individual's mental representation (which, as we have seen in the previous section, is impossible in its entirety) a spatial database can manage and will not confuse these landmarks in conversations with other persons. The following chapter will especially expand this model from a single measure of *landmarkness* to a vector. Such an extension is relatively straight forward and does not change the principles of the model:

```
data Lns = Lns Double Double Double
  – new data type: landmarkness captured by a three-valued vector,
  – for example, by visual, semantic and structural salience (Ch. 5)
```

The extension does not change the principles of the model, only the data string in entities gets longer:

```
Entity ... [Ls Pedestrian (Lns 0.7 0.8 0.4) ...,
    Ls Motorist (Lns 0.9 0.8 0.6 ...]
```

Accordingly, some operations such as *hasLns* need to be adapted as well to cope with the extended data type. Semantically *hasLns* can stay the same. If one of the landmarkness values is larger than 0 it would return *True*. New operations should cater for such a new data type, like computing an overall landmarkness value from a vector of values. The next section will do this by suggesting weighted averages. Then, *hasLns* could be redefined to return *True* if the *overall* landmarkness value is larger than 0.

Chapter 6, finally, will be dedicated to use this model in intelligent spatial communication with persons. Most of this work will be about the selection of appropriate landmark references in machine-generated messages. Here all three data types—the landmarkness context, the landmarkness value, and the preferred reference—will be engaged with. Further selection criteria will be generated by the request itself, which is a part of the communication context that can hardly be anticipated and pre-computed. For example, if someone asks for a route to the train station this particular route will have decision points, and requires references to anchor the orientation decisions to these locations. Hence, selection requires an integration of stored and ad-hoc criteria.

The rest of the book, however, uses a (semi-formal) procedural syntax for specifying algorithms instead of Haskell. The majority of the literature we are compiling and reviewing in the following chapters used procedural syntax already, and while we might edit and homogenize individual styles we preserve to a large extent the original thoughts and approaches.

4.2.3   Formal Models for Elements of the City

As we have seen before each of Lynch's elements can have some *landmarkness*, which varies according to context. Thus, computational models promising to extract elements of the city must be highly relevant for modelling landmarks.

A notable formal spatial modelling stream has grown out of urban research. *Space Syntax* [27] is a way to express the configuration of space by visual accessibility. Graphs constructed from axial lines, which are the longest lines of visibility in an environment, allow a quantitative analysis of configuration properties. Even some aspects of cognition are implicit in space syntax analysis since configuration, and especially the legibility of a configuration is correlated with movement behavior and hence likely to be correlated with mental spatial representations [43]. Despite expected correlations, however, space syntax as such lacks a deeper cognitive grounding.

Nevertheless, measures of space syntax have been used to redefine a subset of Lynch's elements. Some approaches focus on exploring axial lines and isovists in a map of the city [8], other approaches on isovists in a three-dimensional urban digital model [42]. Isovists [2] are the set of all points visible from a given vantage point in space and with respect to a particular environment. Since the isovist is a property of a location the vantage point can be characterized by a measure, e.g., the size of the surface that is visible. Morello and Ratti argue that characterizing location by visibility is about legibility of urban space—Lynch's original interest. Their redefinition of several of Lynch's elements in terms of visibility allows to capture Lynch's elements in a formal, algorithmic way.

Also a link between scaling properties of the city's street network structure and the salient elements of the city form has been suggested [31]. Scaling properties refer here to the typical long tail distribution of elements in the street network by their spatial granularity. For example, a city's street network has a few major roads crossing the city, and many smaller streets refining the network locally. Traffic, or human movement, is similarly distributed with arterial roads being regularly congested, and lighter traffic in the more narrow streets. Clearly there is a link between scaling and hierarchical mental spatial representations. Timpf et al. [55] have even suggested a formal (conceptual) model of human route planning based on such a hierarchy in the street network. They distinguish a planning process happening on a coarse level of the hierarchy (major roads), an instruction or description process involving all roads abstracted to linear objects, and the actual driving process, which involves finer grained levels of a hierarchy, such as lanes and traffic lights. By the way, such a conceptual model explains the correlation between major roads and major traffic.

Tomko and Winter [58] provide a complementary approach to explore the elements of the city formally, providing a means to computationally approximate spatial mental representations of urban environments. Only their model is sensitive to the context of mode of mobility. In their approach objects in the world are classified into Lynch's elements depending on this mode of mobility. Their model is also hierarchic by spatial granularity. The instances of Lynch's elements at each level of granularity are characterized by reference regions. Underlying these reference regions is what Raper called geographic relevance [45], a partition of space such that each cell contains one relevant object. This relevant object would be used as a direct relatum to locate a locatum. It is the object anchoring the location, or the local landmark. The third component of their model are *functional relationships*. These are the relationships between the spatial objects that stem from the actions they afford to a person in some mobility context. For example, for a pedestrian a street is classified as path, and an intersection as node. If they are topologically connected their functional relationship is described by verbs of affordance such as "connect", "follow" or "cross". As Tomko and Winter demonstrate, this model allows to switch between contexts, and to model the acquisition of functional (context-specific) spatial knowledge. This way they suggest a formal model of learning a mental spatial representation by locomotion based on cognitive principles.

4.2.4   Formal Models for Place References

A formal model of landmarks needs linking to other developments of cognitively motivated formal models of spatial objects. One of them, *place*, has been regarded in detail in this book (Sect. 1.2.1). Compatible with our notion of place, one of the earliest advances to capture the notion of place was by affordance [32], recognizing that place is an entity of assigned meaning, or a meaningful configuration, rather than a physical object abstracted to an anchor point. Typical affordances of a place are expressed by the image schemata of *container*, affording shelter, and *surface*, affording movement. The latter has been formalized recently [51]. Approaching place is particularly interesting because it is another fundamental cognitive concept that is missing in spatial information systems [64].

An alternative approach to characterize places can be made by means of contrast [63]. The contrast principle is rather cognitive and linguistic, compared to the perceptual affordance mentioned above. Modelling place by contrast means explaining the meaning of a (reference to a) place by a contrast set, which also captures a communication context. For example, *Alexanderplatz* is a reference to a place in Berlin. Talking to a tourist about a local attraction the reference can mean *Alexanderplatz* in contrast to *Unter den Linden*, *Pergamonmuseum* and *Kurfürstendamm*. Talking to a nearby friend on the phone to negotiate a meeting point the reference can mean *Alexanderplatz* in contrast to *Spandauer Strasse*, *Karl-Liebknecht-Strasse*, and *Karl-Marx-Allee*. The contrast principle allows explaining the localization of objects whilst avoiding geometrically specified positions in space or characterizing places by boundaries. Places are described as prototypes or centres, or even dimensionless entities in information space. The spatial extent of a place is only a refinement, but not core of the meaning of place. This approach fits well to proximity relations between elements. Proximity to dimensionless entities is cognitively and computationally more directly accessible than distances between spatially extended objects, which require more complex concepts such as the Hausdorff distance [25].

Contrast sets always provide a key to choose an appropriate level of spatial granularity for reference regions. However, in a communication the current contrast set is only sometimes explicit. In other communication situations the contrast set has to be concluded from factors such as the location of the person, other locative expressions, or activities mentioned.

4.2.5 Formal Models for Hierarchies

Two types of hierarchies among landmarks were identified in human spatial reasoning and communication, *spatial granularity* and cognitive *salience*. In order to enable a machine to generate cognitively efficient expressions in human-computer interaction—expressions that zoom in or out—both kinds of hierarchies need to be available.

4.2.5.1 Hierarchies and Reference Regions

Configurations of objects, such as the landmarks in our thought experiment in Sect. 2, consist of a set of geographic (environmental) objects that typically satisfy to be non-overlapping, littering the otherwise empty environmental space, and stationary. The entities representing these objects shall be modelled as conceptual nodes in information space. Their spatial extent might be represented as a further property, but is of no relevance for this model. Furthermore this approach facilitates efficient proximity relations between elements. For modelling proximity we resort to Delaunay triangulations and their dual, Voronoi diagrams [1, 3, 10, 60, 61]. This way connections between objects are by design limited to nearest neighbors, which is not necessarily the case in mental spatial representations. While the actual characteristic of relationships in mental spatial relationships remains opaque, the general principle of cognitive efficiency would at least support such a design choice as a good approximation. Theoretical evidence assuming that nearest objects are (at least stronger) linked in mental spatial representations exists. For example, human estimations of distances and directions are better for nearby objects [39]. Also, wayfinding and other spatial activities seem to be easier for parts of the environment that are nearby and familiar: "For an individual there is not an equal probability of behavior occurring in all sections of the environment. There is a spatial bias for behavior close to his residence owing to least effort considerations" ([6], p. 368). That things near to each other are more correlated than distant things was an observation made originally by Tobler for geographic space [56], and later also confirmed as a mental default assumption [41]. As a design choice Voronoi diagrams provide an unambiguous set of relationships[5] and an efficient reasoning on a small number of links between likely correlated objects.

A number of properties follow from this design choice:

- Objects in one configuration can belong to different object classes (e.g., the variety of objects within a city), but they must be spatially separable (non-overlapping), i.e., from the same level in a hierarchy of spatial granularities. For example, one configuration can consist of cities (in the context of representing the geographic space of a country), and another configuration can consist of salient buildings within a city (in the context of representing the geographic space of the city), but a configuration cannot contain a city and a building of that city. The city is an agglomerate of buildings (hence, the building is contained by the city), and accordingly, city and buildings belong to different levels of spatial granularity.
- The objects of a configuration form a context via contrast [63]. The spatial meaning of a locative expression can be specified by its current contrast set. For example, the locative expression "[at / pass] the church" can have a meaning in a configuration of {(this) church, city hall, museum, cinema}, and it would have another spatial meaning in {(this)church, the other church north of it, and the third church south of it}.

The Voronoi diagram associates always the nearest landmark for a localization task. "I am near the church" describes my location anchored by the church (in some appropriate contrast set). "I am in Zurich" anchors my location at another level of granularity and also evokes a different contrast set.

Choosing Voronoi diagrams is of course a simplifying heuristic, which brings in uncertainty at two levels:

- The application of a Voronoi diagram is a (heuristic) simplification of a more complex geographic reality. For example, a configuration of all European capitals provides a stable Voronoi diagram to locate other objects in this reference frame. However, the Voronoi diagram is only a poor reflection of the nations' boundaries and hence may produce non-preferred nearness relationships.

---

[5]Mathematically there are exceptions: relationships become ambiguous when nearest neighbors are arranged on a line or in a rectangle. But practically this nearly never happens in geographic space.

- The Voronoi diagram assumes equally-weighted objects, but according to observations like the ones by Sadalla et al. distances are perceived asymmetrically. A better reflection of neighborhoods of weighted objects would be a weighted Voronoi diagram.

4.2.5.2 Hierarchies by Spatial Granularity

The hierarchy by spatial granularity is typically part of the data structures of a spatial database. Such hierarchic structures exist for gazetteers, which can link each entity to the larger object it belongs to [26], as well as geometric spatial databases, which provide the topological structures between layers of object categories in order to reason for containment. Here, we refer to models that actually make use of these hierarchic structures in either generating or understanding place descriptions [57, 68].

4.2.5.3 Hierarchies by Cognitive Salience

Hierarchy by cognitive salience cannot be derived from database content. It depends on the partial order that is imposed by cognitive and social properties such as prominence or uniqueness. Prominence imposes a partial order between objects by ranking them for being known by a population. A partial order by uniqueness can be established by the size of the region in which an object is unique. For example, the Royal Melbourne Hospital is unique in Melbourne, and the corner store opposite of the hospital is unique within the vista space of the hospital, but not within the city. Alternatively, a partial order can be produced directly from the cognitive and social properties of the objects. These properties, we have seen before, may be outstanding only to some degree, but not be unique. Sorrows and Hirtle [53] considered visible, structural and semantic properties of objects as relevant in this context, and others have developed a *measure of salience* based on these properties [33, 46].

The partial order is useful to establish a leveled hierarchy. The quantitative prominence or salience measures can be linked together with the spatial neighborhood structure of the reference regions to define these levels [66]. The process starts with the Voronoi diagram of all known landmarks. Each landmark's reference region is the area where the dominance of this landmark is stronger than the dominance of any other competing landmark, but with the high density of landmarks these reference regions will be relatively small even for very prominent landmarks. Thus, in an recursive procedure the quantity of landmarkness is considered for each cell, such that the local maxima are lifted to the next level of the hierarchy, and a new Voronoi diagram is computed. This way the reference regions grow from level to level, as the seed elements of the Voronoi diagram are focusing on the more salient ones.

For example, Fig. 4.1 shows the Voronoi Diagram of the set of Melbourne's inner city train stations *Melbourne Central, Southern Cross, Flagstaff, Parliament,* and *Flinders Street.* Considering that these train stations have largely varying landmarkness, proportional to the traffic they attract, a partial order can be established. The

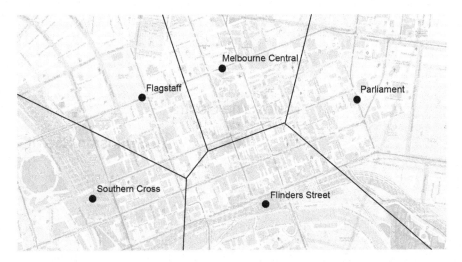

Fig. 4.1 The Voronoi diagram of the set of Melbourne's inner city train stations Southern Cross, Flagstaff, Melbourne Central, Parliament, and Flinders Street. Map copyright OpenStreetMap contributors, used under CC BY-SA 2.0

most central station in the public transport network is Flinders Street Station, so in the process of establishing a next level of the salience hierarchy the only local maximum, Flinders Street Station, gets lifted. Thus the final salience hierarchy consists of a root node (Flinders Street Station), and a second level, containing all stations. This hierarchy reflects that in some communication contexts "the train station" refers to Flinders Street Station, city-wide. In other communication contexts the contrast set may consist of all five stations, and a finer distinction has to be made.

These salience hierarchies are also suited to map hierarchical cognitive reasoning across hierarchies of spatial granularity (Sect. 4.2.5.2). In order to illustrate this, let us consider a hierarchical verbal place description. "I am at Melbourne Central, Swanston Street exit" is a description zooming in, using landmarks from different levels of spatial granularity as relata. In Voronoi Diagrams this zooming behavior can be reflected. The first relatum, *Melbourne Central* may be understood in the context of a train traveller, i.e., applying the contrast set of Fig. 4.1. The corresponding Voronoi cell was perceived by the speaker to be too coarse for this particular communication purpose, and a refinement was sought. Figure 4.2 shows a local refinement of the original Voronoi diagram built from the salient elements of Melbourne Central—the new contrast set. Without a commitment to a geometrically exact position the speaker conveys a location closer to the *Swanston Street exit* than to any other salient element of the train station.

Hence, spatial entities bearing landmarkness information and being explicitly or implicitly organized in hierarchical data structures, can reflect hierarchical cognitive reasoning.

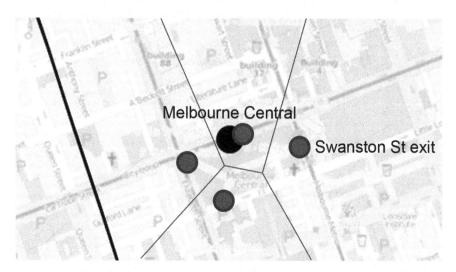

Fig. 4.2 A switch in spatial granularity from stations to exits (or other elements) of stations. Map copyright OpenStreetMap contributors, used under CC BY-SA 2.0

4.3   Landmarks in Geographic Data Structures

Our conceptual model requires data models capable to represent the landmarkness of entities of any spatial granularity while representing this landmarkness in a context-dependent way. In this section we will review existing data models for their compliance, and discuss shortfalls in particular cases.

4.3.1   POI Data: Substitutes for Landmarks

Popular web mapping services, car navigation services or mobile location-based services have internal data structures for storing points of interests (POI), and accordingly also standards for the description and exchange of POIs exist.[6] We have discussed already the fundamental differences between landmarks and POIs, but here we come back to POI since they appear to be used in lack of the more appropriately suited landmarks in service-generated place and route descriptions.

POIs are stored as traditional Geographic Information System layers containing categories of geographic objects such as places of interest, businesses, transit stations, mountains, parks and golf courses. This arbitrary mixture of categories, by the way, are just the default layers taken from a prominent search engine's webmapping application. These layers can also be expanded by further categories,

---

[6]http://www.w3.org/2010/POI/, http://openpoi.ogcnetwork.net/, last visited 3/1/2014.

even for a layer of user generated POIs. Storing these categories as layers enables the user to switch them on or off. For the service itself all objects in a layer are equal, and switching on or off by users means the service itself has no concept of relevance, or mechanism to capture and consider a communication context. The original statement "We've found it super useful for checking out what's nearby a hotel we'll be staying at, orienting ourselves, getting the feel for a neighborhood, or just browsing around for fun"[7] makes clear that the purpose of POI is quite different from landmarks, and far from the intelligent communication capabilities of the machine we are after. This approach to POI is also more prone to commercial interference such as sponsoring.

4.3.2 Landmarks in OpenStreetMap

Since spatial databases do not have a type *landmark* in their taxonomies, as landmarkness is rather a property of existing entities than a type, user generated content is a tempting approach to collect the necessary data. This temptation relies on two assumptions:

- Landmark data can be collected from scratch, and can be acquired in high density. Only crowds have shown to be that agile.
- Landmarks are cognitive anchor points, i.e., everybody has an intuitive understanding of what a landmark is, and for which purpose a particular local embodied experience of an environment might be a suited reference point. Crowds formed by people in-situ can report these experiences [18].

User-generated content has been defined loosely as content that "comes from regular people who voluntarily contribute data, information, or media that then appears before others in a useful or entertaining way, usually on the Web–for example, restaurant ratings, wikis, and videos" ([35], p. 10). OpenStreetMap[8] is a prime example of user-generated content in general, and especially as user generated *spatial* content—some people speak of *volunteered geographic information* [18]— OpenStreetMap is a world-wide success story [22, 23]. OpenStreetMap is a project maintaining a platform to create a set of map data that is open, free to use, and editable for everyone. OpenStreetMap is based on a peer production model, similar to Wikipedia. The project was founded in 2004, and had reached one million registered contributors before 2013.

Peer production means that volunteers (registered contributors) upload geographic information to a central database in the cloud, which can be accessed by anyone, and edited by registered contributors. Registration is free. Once a

---

[7]http://google-latlong.blogspot.com.au/2009/08/i-didnt-know-that-was-there.html, last visited 3/1/2014.

[8]http://www.openstreetmap.org, last visited 3/1/2014.

geographic object has been created in OpenStreetMap, it can be tagged by additional information. This latter property offers the prerequisite to implement our formal landmark model presented in Sect. 4.2.2 by tagging objects with context-dependent properties of landmarkness. According to the model above, a tag should store at least a name, a context, and measures of landmarkness.

Tags are key-value pairs describing properties of data types of all dimensions, nodes, lines or areas. OpenStreetMap comes with a long list of conventionalized tags,[9] but users can add further tags. Such further tags, however, need then also to be introduced in the mapping and navigation services based on OpenStreetMap. For each new tag a community has to be ready to use them, and to use them consistently. Without community support, tag use will remain local and probably fade out over time.

It is straight forward to define key-value pairs to represent the landmarkness of the objects according to the above formal model (Sect. 4.2.2) in OpenStreetMap, and it is also straight forward to develop mobile applications that support the community in capturing this landmark information [17,49] and using this landmark information in services. In lack of properly defined key-value pairs no community-accepted and supported landmark tag has been introduced in OpenStreetMap so far.

4.3.3 Landmarks in OpenLS

The closest contender for a suited data model is probably an inclusion of landmarks in OpenLS [24].

OpenLS is an international standard by ISO and OGC.[10] OpenLS describes an open platform for location-based services. Among its core services is a route service. OpenLS specifies primarily the interaction between client and server and the format in which the transferred data is encoded. The role of a route service in OpenLS is to determine a travel route between two points and to collect the navigation information required to communicate this route.

For the interaction between client and server OpenLS comes with an XML schema for location services, called XLS. XLS is an exchange data model for route services, this means for structuring and encoding requests and responses regarding route descriptions. XLS only caters for POI, and does not yet cater for an abstract data type *landmark*. However, XLS offers explicitly an optional attribute, which can serve different purposes, among others also describing a landmark. This attribute allows for providing the name of the landmark and its location with respect of the side of the route. Other information about the landmark cannot be encoded in this attribute, such that information necessary for the cognitive ergonomic use of this landmark cannot be encoded and is lost.

---

[9]http://wiki.openstreetmap.org/wiki/Map_Features, last visited 3/1/2014.

[10]http://www.opengeospatial.org/standards/ols, last visited 3/1/2014.

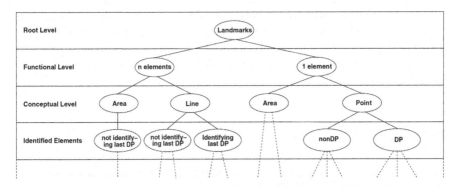

Fig. 4.3 The top levels of a landmark taxonomy along routes (from [24])

Therefore, Hansen et al. [24] suggest an extension of XLS to capture the cognitively meaningful semantics of landmarks. Based on a taxonomy shown in Fig. 4.3 they develop a hierarchic abstract data type for landmarks. Many of these distinctions will become clearer in Chap. 5, where we will be dealing with selecting landmarks along routes. But the taxonomy shows already that these characteristics of landmarks are route-dependent and thus not suited to be captured in the conceptual model above. They can only be determined in an ad-hoc manner for a particular conversation or route request.

Later this work was extended to enable spatial chunking, the cognitively based grouping of consecutive decision points to achieve cognitively ergonomic route descriptions [34]. Route descriptions are called here cognitively ergonomic if they adhere to the cognitive principles we have collected so far. In short, they are route instructions that consider the prior knowledge of the traveller, the structure of the environment, and also the human cognitive capacities. These criteria provide the scope for flexibly adapting the granularity of route instructions, and for chunking. Klippel et al. call their extension an *Urban Knowledge Data Structure* (UKDS), adding chunking types to the representation of elementary instructions with landmarks.

4.4 Summary

In this section we have presented a formal model suitable for developing entities in spatial databases that can represent some landmarkness of the represented objects in certain contexts. We have also discussed variations of this model, including structured context models, vectors characterizing an objects' landmarkness in multiple dimensions, and the landmarkness of types.

What we have not yet discussed is the computational semantics of landmarkness values. Formally, they represent fuzzy membership values to a vaguely defined category *landmark* (in a certain context), but questions whether this landmarkness comes out of people's rating, data mining in harvested user generated data, or expert assignment, and whether it represents an object's prominence, its cognitive salience, or its centrality in an environment have been left out so far. They will be discussed in the next chapter.

References

1. Aurenhammer, F.: Voronoi diagrams: A survey of a fundamental geometric data structure. ACM Comput. Surv. **23**(3), 345–405 (1991)
2. Benedikt, M.L.: To take hold of space: Isovist and isovist fields. Environ. Plan. B **6**(1), 47–65 (1979)
3. de Berg, M., van Kreveld, M., Overmars, M., Schwarzkopf, O.: Computational Geometry, 2 edn. Springer, Berlin (2000)
4. Billen, R., Van de Weghe, N.: Qualitative spatial reasoning. In: Thrift, N., Kitchin, R.M. (eds.) International Encyclopaedia of Human Geography, pp. 12–18. Elsevier, Amsterdam, NL (2009)
5. Bittner, T., Frank, A.U.: On the design of formal theories of geographic space. J. Geogr. Syst. **1**(3), 237–275 (1999)
6. Briggs, R.: Urban cognitive distance. In: Downs, R.M., Stea, D. (eds.) Image & Environment, pp. 361–388. Aldine Publishing Company, Chicago (1973)
7. Cohn, A.G., Renz, J.: Qualitative spatial representation and reasoning. In: van Harmelen, F., Lifschitz, V., Porter, B. (eds.) Handbook of Knowledge Representation, pp. 551–596. Elsevier, Amsterdam, NL (2008)
8. Conroy Dalton, R., Bafna, S.: The syntactical image of the city: A reciprocal definition of spatial elements and spatial syntaxes. In: Hanson, J. (ed.) Fourth International Symposium on Space Syntax, pp. 59.1–59.21. London, (2003)
9. Dey, A.K.: Understanding and using context. Personal Ubiquitous Comput. **5**(1), 4–7 (2001)
10. Edelsbrunner, H.: Algorithms in Combinatorial Geometry. In: EATCS Monographs on Theoretical Computer Science, vol. 10. Springer, Berlin (1987)
11. Frank, A.U.: Formal models for cognition: Taxonomy of spatial location description and frames of reference. In: Freksa, C., Habel, C., Wender, K.F. (eds.) Spatial Cognition. Lecture Notes in Artificial Intelligence, vol. 1404, pp. 293–312. Springer, Berlin (1998)
12. Frank, A.U.: One step up the abstraction ladder: Combining algebras – From functional pieces to a whole. In: Freksa, C., Mark, D.M. (eds.) Spatial Information Theory. Lecture Notes in Computer Science, vol. 1661, pp. 95–107. Springer, Berlin (1999)
13. Frank, A.U., Bittner, S., Raubal, M.: Spatial and cognitive simulation with multi-agent systems. In: Montello, D.R. (ed.) Spatial Information Theory, Lecture Notes in Computer Science, vol. 2205, pp. 124–139. Springer, Berlin (2001)
14. Frank, A.U., Kuhn, W.: Specifying open GIS with functional languages. In: Egenhofer, M.J., Herring, J.R. (eds.) Advances in Spatial Databases. Lecture Notes in Computer Science, vol. 951, pp. 184–195. Springer, Berlin (1995)
15. Frank, A.U., Kuhn, W.: A specification language for interoperable GIS. In: Goodchild, M.F., Egenhofer, M., Fegeas, R., Kottman, C. (eds.) Interoperating Geographic Information Systems, pp. 123–132. Kluwer, Norwell (1999)
16. Gangemi, A., Guarino, N., Masolo, C., Oltramari, A., Schneider, L.: Sweetening ontologies with DOLCE. In: Benjamins, V.R. (ed.) Knowledge Engineering and Knowledge Management: Ontologies and the Semantic Web. Lecture Notes in Artificial Intelligence, vol. 2473, pp. 166–181. Springer, Berlin (2002)

17. Ghasemi, M., Richter, K.F., Winter, S.: Landmarks in OSM. In: 5th Annual International OpenStreetMap Conference. Denver (2011)
18. Goodchild, M.: Citizens as sensors: The world of volunteered geography. GeoJournal **69**(4), 211–221 (2007)
19. Grenon, P., Smith, B.: Snap and span: Towards dynamic spatial ontology. Spat. Cogn. Comput. **4**(1), 69–104 (2004)
20. Gruber, T.R.: Toward principles for the design of ontologies used for knowledge sharing. Int. J. Hum. Comput. Stud. **43**(5–6), 907–928 (1995)
21. Guarino, N., Oberle, D., Staab, S.: What is an ontology? In: Staab, S., Studer, R. (eds.) Handbook on Ontologies, 2nd edn., pp. 1–17. Springer, Dordrecht (2009)
22. Haklay, M.: How good is volunteered geographic information? A comparative study of OpenStreetMap and Ordnance Survey datasets. Environ. Plan. B **37**(4), 682–703 (2010)
23. Haklay, M., Weber, P.: OpenStreetMap: User-generated street maps. Pervasive Comput. **7**(4), 12–18 (2008)
24. Hansen, S., Richter, K.F., Klippel, A.: Landmarks in OpenLS - a data structure for cognitive ergonomic route directions. In: Raubal, M., Miller, H., Frank, A.U., Goodchild, M.F. (eds.) Geographic Information Science. Lecture Notes in Computer Science, vol. 4197, pp. 128–144. Springer, Berlin (2006)
25. Hausdorff, F.: Grundzüge der Mengenlehre. Veit & Company, Leipzig (1914)
26. Hill, L.L.: Georeferencing: The Geographic Associations of Information. Digital Libraries and Electronic Publishing. MIT Press, Cambridge (2006)
27. Hillier, B., Hanson, J.: The Social Logic of Space. Cambridge University Press, Cambridge (1984)
28. Janelle, D.G.: Impact of information technologies. In: Hanson, S., Giuliano, G. (eds.) The Geography of Urban Transportation, pp. 86–112. Guilford Press, New York (2004)
29. Janowicz, K., Keßler, C.: The role of ontology in improving gazetteer interaction. Int. J. Geogr. Inf. Sci. **22**(10), 1129–1157 (2008)
30. Janowicz, K., Raubal, M.: Affordance-based similarity measurement for entity types. In: Winter, S., Duckham, M., Kulik, L., Kuipers, B. (eds.) Spatial Information Theory. Lecture Notes in Computer Science, vol. 4736, pp. 133–151. Springer, Berlin (2007)
31. Jiang, B.: Computing the image of the city. In: Campagna, M., De Montis, A., Isola, F., Lai, S., Pira, C., Zoppi C. (eds.) Proceedings of the 7th International Conference on Informatics and Urban and Regional Planning, pp. 111–121. Cagliari, Italy (2012)
32. Jordan, T., Raubal, M., Gartrell, B., Egenhofer, M.J.: An affordance-based model of place in GIS. In: Poiker, T.K., Chrisman, N. (eds.) 8th International Symposium on Spatial Data Handling, pp. 98–109. IGU, Vancouver, Canada (1998)
33. Klippel, A., Winter, S.: Structural salience of landmarks for route directions. In: Cohn, A.G., Mark, D.M. (eds.) Spatial Information Theory. Lecture Notes in Computer Science, vol. 3693, pp. 347–362. Springer, Berlin (2005)
34. Klippel, A., Hansen, S., Richter, K.F., Winter, S.: Urban granularities - a data structure for cognitively ergonomic route directions. GeoInformatica **13**(2), 223–247 (2009)
35. Krumm, J., Davies, N., Narayanaswami, C.: User generated content. Pervasive Comput. **7**(4), 10–11 (2008)
36. Kuhn, W.: Core concepts of spatial information for transdisciplinary research. Int. J. Geogr. Inf. Sci. **26**(12), 2267–2276 (2012)
37. Maling, D.H.: Coordinate systems and map projections for GIS. In: D. Maguire, M. Goodchild, D. Rhind (eds.) Geographical Information Systems: Principles and Applications, pp. 135–146. Longmans Publishing Co. (1991)
38. Maturana, H.R.: Neurophysiology of cognition. In: Garvin, P. (ed.) Cogntion: A Multiple View, pp. 3–23. Spartan Books, New York (1970)
39. McNamara, T.P.: Mental representations of spatial relations. Cogn. Psychol. **18**(1), 87–121 (1986)
40. Molenaar, M.: An Introduction to the Theory of Spatial Object Modelling for GIS. Research Monographs in Geographic Information Systems. Taylor and Francis, London (1998)

41. Montello, D.R., Fabrikant, S.I., Ruocco, M., Middleton, R.S.: Testing the first law of cognitive geography on point-display spatializations. In: Kuhn, W., Worboys, M.F., Timpf, S. (eds.) Spatial Information Theory. Lecture Notes in Computer Science, vol. 2825, pp. 316–331. Springer, Berlin (2003)
42. Morello, E., Ratti, C.: A digital image of the city: 3D isovists in Lynch's urban analysis. Environ. Plan. B **36**(5), 837–853 (2009)
43. Penn, A.: Space syntax and spatial cognition: Or why the axial line? Environ. Behav. **35**(1), 30–65 (2003)
44. Purves, R.S.: Methods, examples, and pitfalls in the exploitation of the geospatial web. In: Hesse-Biber, S.N. (ed.) The Handbook of Emergent Technologies in Social Research, pp. 592–622. Oxford University Press, New York (2011)
45. Raper, J.: Geographic relevance. J. Doc. **63**(6), 836–852 (2007)
46. Raubal, M., Winter, S.: Enriching wayfinding instructions with local landmarks. In: Egenhofer, M.J., Mark, D.M. (eds.) Geographic Information Science, Lecture Notes in Computer Science, vol. 2478, pp. 243–259. Springer, Berlin (2002)
47. Renz, J.: Qualitative Spatial Reasoning with Topological Information, Lecture Notes in Computer Science, vol. 2293. Springer, Berlin (2002)
48. Richter, K.F., Tomko, M., Winter, S.: A dialog-driven process of generating route directions. Comput. Environ. Urban Syst. **32**(3) (2008)
49. Richter, K.F., Winter, S.: Harvesting user-generated content for semantic spatial information: The case of landmarks in OpenStreetMap. In: Hock, B. (ed.) Proceedings of the Surveying and Spatial Sciences Biennial Conference 2011, pp. 75–86. Surveying and Spatial Sciences Institute, Wellington (2011)
50. Sadalla, E.K., Burroughs, J., Staplin, L.J.: Reference points in spatial cognition. J. Exp. Psychol. Hum. Learn. Mem. **6**(5), 516–528 (1980)
51. Scheider, S., Kuhn, W.: Affordance-based categorization of road network data using a grounded theory of channel networks. Int. J. Geogr. Inf. Sci. **24**(8), 1249–1267 (2010)
52. Smith, B.: Beyond concepts: Ontology as reality representation. In: Varzi, A.C., Vieu, L. (eds.) Third International Conference on Formal Ontology and Information Systems (FOIS), pp. 73–84. IOS Press, Turin (2004)
53. Sorrows, M.E., Hirtle, S.C.: The nature of landmarks for real and electronic spaces. In: Freksa, C., Mark, D.M. (eds.) Spatial Information Theory. Lecture Notes in Computer Science, vol. 1661, pp. 37–50. Springer, Berlin (1999)
54. Studer, R., Benjamins, V.R., Fensel, D.: Knowledge engineering: Principles and methods. Data Knowl. Eng. **25**(1–2), 161–197 (1998)
55. Timpf, S., Volta, G.S., Pollock, D.W., Frank, A.U., Egenhofer, M.J.: A conceptual model of wayfinding using multiple levels of abstraction. In: Frank, A.U., Campari, I., Formentini, U. (eds.) Theories and Methods of Spatio-Temporal Reasoning in Geographic Space. Lecture Notes in Computer Science, vol. 639, pp. 348–367. Springer, Berlin (1992)
56. Tobler, W.: A computer movie simulating urban growth in the Detroit region. Econ. Geogr. **46**(2), 234–240 (1970)
57. Tomko, M., Winter, S.: Pragmatic construction of destination descriptions for urban environments. Spat. Cogn. Comput. **9**(1), 1–29 (2009)
58. Tomko, M., Winter, S.: Describing the functional spatial structure of urban environments. Comput. Environ. Urban Syst. **41**(September), 177–187 (2013)
59. Varela, F.J.: Whence perceptual meaning? A cartography of current ideas. In: Varela, F.J., Dupuy, J.P. (eds.) Understanding Origins, Boston Studies in the Philosophy and History of Science, vol. 130, pp. 235–263. Springer, Amsterdam (1992)
60. Voronoi, G.: Nouvelles applications des paramètres continus à la théorie des formes quadratiques (first part). Journal für die Reine und Angewandte Mathematik (Crelle's Journal) **1908**(133), 97–178 (1908)
61. Voronoi, G.: Nouvelles applications des paramètres continus à la théorie des formes quadratiques (second part). Journal für die Reine und Angewandte Mathematik (Crelle's Journal) **1909**(134), 67–182 (1909)

62. Wallgrün, J.O., Frommberger, L., Wolter, D., Dylla, F., Freksa, C.: A toolbox for qualitative spatial representation and reasoning. In: Barkowsky, T., Knauff, M., Ligozat, G., Montello, D.R. (eds.) Spatial Cognition V. Lecture Notes in Artificial Intelligence, vol. 4387, pp. 39–58. Springer, Berlin (2007)
63. Winter, S., Freksa, C.: Approaching the notion of place by contrast. J. Spat. Inf. Sci. **2012**(5), 31–50 (2012)
64. Winter, S., Kuhn, W., Krüger, A.: Does place have a place in geographic information science? Spat. Cogn. Comput. **9**(3), 171–173 (2009)
65. Winter, S., Nittel, S.: Formal information modeling for standardisation in the spatial domain. Int. J. Geogr. Inf. Sci. **17**(8), 721–741 (2003)
66. Winter, S., Tomko, M., Elias, B., Sester, M.: Landmark hierarchies in context. Environ. Plann. B Plann. Des. **35**(3), 381–398 (2008)
67. Winter, S., Truelove, M.: Talking about place where it matters. In: Raubal, M., Mark, D.M., Frank, A.U. (eds.) Cognitive and Linguistic Aspects of Geographic Space: New Perspectives on Geographic Information Research. Lecture Notes in Geoinformation and Cartography, pp. 121–139. Springer, Berlin (2013)
68. Wu, Y., Winter, S.: Interpreting destination descriptions in a cognitive way. In: Hois, J., Ross, R., Kelleher, J., Bateman, J. (eds.) Workshop on Computational Models for Spatial Language Interpretation and Generation (CoSLI-2), vol. 759, pp. 8–15. CEUR-WS.org, Boston (2011)
69. Zadeh, L.A.: Fuzzy sets. Inf. Control **8**, 338–353 (1965)
70. Zheng, Y., Zhou, X. (eds.): Computing with Spatial Trajectories. Springer, New York (2011)

Chapter 5
Computational Aspects: How Landmarks Can Be Observed, Stored, and Analysed

Abstract In this chapter, we will explore how to 'compute' a landmark. We will look at ways to calculate that some geographic object sticks out from its background. We will also discuss approaches for selecting the most appropriate landmark for describing specific spatial situations. Both these aspects are important steps for the integration of landmarks in computational services. Therefore, in a third part of this chapter we will discuss commonalities and differences between both aspects, where and why the presented approaches may fail, and what alternatives there are for overcoming these shortcomings.

5.1 Computing a Landmark

The previous chapters have established the importance of landmarks for our understanding of an environment. We have also highlighted how this impacts on our way of communicating. Furthermore, Chap. 4 has shown that it is possible to capture at least the principle aspects of landmarks and landmarkness in a formal way, making them accessible to computers. This chapter will discuss how computers may be able to populate these formal specifications. This will encompass approaches of determining whether and how a geographic object sticks out from the background, as well as how to select the most appropriate landmark in a given situation. Both are important aspects of integrating landmarks into geospatial services, however, they are often treated separately.

To provide an initial idea of the differences between both aspects, consider the situation depicted in Fig. 5.1. In the situation on the left, some algorithm determined for each geographic object whether it fulfills the criteria of landmarkness as discussed in the last chapter. In particular, such an algorithm identifies those objects that are *salient* in their local surroundings. The figure in (a) shows all objects of those contained in the grey areas that the algorithm considered to be salient. This results in a set of *landmark candidates*, which form the input for other algorithms that select the most *relevant* landmark for a given situation. For example,

K.-F. Richter and S. Winter, *Landmarks: GIScience for Intelligent Services*,
DOI 10.1007/978-3-319-05732-3__5, © Springer International Publishing Switzerland 2014

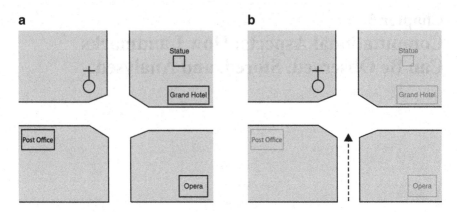

Fig. 5.1 Computing a landmark: (**a**) identifying geographic objects that may serve as a landmark in principle; (**b**) selecting the most suited landmark for a specific situation

in Fig. 5.1b, in order to describe the marketplace when coming from the south, selecting a landmark candidate that is visible early on is a sensible choice; here, this may be the church.

We term these steps *landmark identification* and *landmark integration*, respectively. They are important steps in computing a landmark. However, they are often performed by different algorithms and research addresses either one or the other. Accordingly, the next section will present approaches to *identifying* landmarks. Section 5.3 then will illustrate approaches to *integrating* landmarks. Section 5.4 will compare the approaches to *identification* and *integration* of landmarks, which will lead to a criticism of existing approaches, presented in Sect. 5.5. That section will also argue for some extensions and alternative approaches, respectively, to circumvent some of the inherent drawbacks that come with the current approaches to computing landmarks. To a large extent it seems these drawbacks are mainly responsible for why landmarks are not used (more) in today's location-based services.

5.2 Landmark Identification

The aim of *landmark identification* is to find *all* geographic objects in a given region that may serve as a landmark *in principle*. To achieve this, for each object its salience for people needs to be determined. Salience can either be computed or inferred from how these objects are referred to in some data source. In other words, algorithms for landmark identification either use attribute data of geographic objects (Sect. 5.2.1) or they mine other document sources to determine an object's salience by the way (and number of times) documents refer to that object (Sect. 5.2.2).

5.2.1 Computing Salience

The first approach to the automatic identification of landmarks has been presented by Raubal and Winter [35]. Their approach reflects the three landmark characteristics of Sorrows and Hirtle [45] discussed in Chaps. 3 and 4. Raubal and Winter's approach aims at capturing perceptual and cognitive aspects of geographic objects, which are then used to calculate landmark salience. Their formal model of landmark salience is based on the concept of *attractiveness*, which reflects the 'landmarkness' of an object, i.e., how strong a landmark candidate it is.

In accordance with Sorrows and Hirtle's conceptual classification of landmark aspects, Raubal and Winter define three different kinds of attractiveness: *visual, semantic,* and *structural.* Geographic objects are visually attractive if they are in sharp contrast to their surroundings or have a prominent spatial location. The formal model for landmark salience includes four measures of visual attractiveness:

- Façade area: If the façade area of an object is significantly larger or smaller than those of the surrounding objects this object becomes well noticeable. In the original model, façade area is simply measured as the product of width and height, assuming rectangular buildings, however, this can be extended to account for more irregular shapes as well.
- Shape: Unusual shapes, especially among more regular, box-like objects, are highly remarkable. Indeed, architects use this to make buildings stand out. The model distinguishes two aspects of shape, the *shape factor* and the *deviation.* The shape factor simply is the proportion of height and width. The deviation is the ratio of the area of the minimum-bounding rectangle (mbr) of the object's façade and its façade area. Again, more complicated measures could be defined if needed.
- Color: For humans, color is a clear indicator of visual attractiveness. For example, a red building will be highly visible among a row of grey buildings. In the model, color is measured as a decimal value derived from the RGB color space.
- Visibility: Like color, visibility is highly important for visual attractiveness. The model assumes a two-dimensional visibility, defined by recognizability within a buffer zone. Visibility is measured as the fraction (over its total size) of a building's front (its façade) in this buffer zone.

While color seems to be an obvious measure for salience, in practice challenges arise for automatically determining color differences in an image because of the *grey world assumption* [3]. Essentially, this assumption states that given enough color variation in an image, the average value of RGB will be a common grey value. As a consequence, color variation may be far less obvious for a computer algorithm than it is for the human eye.

Table 5.1 Properties and measurement of visual, semantic, and structural attraction

Measure	Property	Measurement	Attractiveness Measure
Visual Attraction	Façade	$p_{vf} = \int x \,\vert\, x \in$ façade	$s_{vis} = (p_{vf} +$
	Shape	$p_{vsf} =$ height/width	$p_{vsf} + p_{vsd} +$
		$p_{vsd} = ($area mbr $- \alpha)/$area mbr	$pvc + p_{vv})/5$
	Color	$p_{vc} = [R, G, B]$	
	Visibility	$p_{vv} = \sum x \,\vert\, x$ visible	
Semantic Attraction	Cultural and Historic	$p_{sec} = [0, 1]$	$s_{sem} =$ $(p_{sec} + p_{sem})/2$
	Explicit Mark	$p_{sem} = [0, 1]$	
Structural Attraction	Nodes	$p_{stn} = i + o$	$s_{str} =$
	Boundaries	$p_{stb} =$ cell size $*$ form factor	$(p_{stn} + p_{stb})/2$

Adapted from [35]

Semantic attraction is made up of two measures:

- Cultural and historic importance: This property reflects whether an object is culturally or historically important and, thus, becomes attractive. In the simplest case, this is a Boolean value of true or false, but it may be refined, for example, by measuring importance on a scale of 1 to 5.
- Explicit marks: An object may have explicit marks, such as signs on the front of a building. These signs explicitly label an object communicating its semantics. Again, this property is measured by a Boolean value.

Structural attraction is also measured using two properties:

- Nodes: here, nodes are the intersections in a travel network. The structural importance of a node is defined by the number of incoming and outgoing edges (the node degree). This could be further weighted by accounting for the importance of these edges, for example, ranking highways higher than footpaths.
- Boundaries: boundaries separate two or more areas in geographic space. The structural attractiveness of a boundary is linked to the effort that is required to cross it. The higher its resistance, the higher its attractiveness. However, recent findings show that this may be too simplistic an assumption [48]. The boundary measure is the product of cell size and form factor in the formal model.

Table 5.1 summarizes the properties and measurement of the different kinds of attractiveness used in Raubal and Winter's formal model.

The three individual attractiveness measures need to be combined in order to calculate an overall attractiveness of a geographic object, which then gives the landmark salience. Raubal and Winter's model uses a weighted sum to this end (see Eq. 5.1). The different weights can be adapted to reflect varying importance of the individual attractiveness measures in different contexts, user preferences, or applications. The original model sets all weights as equal, which is sensible if no further information about each measure's influence is known.

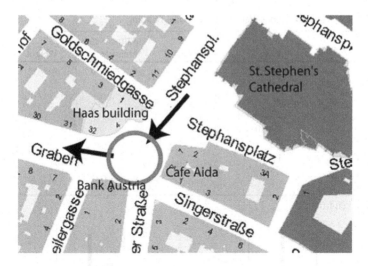

Fig. 5.2 Example of a wayfinding situation where a reference to a landmark may be used to describe the required action; adapted from [35]

$$s = w_{vis} * s_{vis} + w_{sem} * s_{sem} + w_{str} * s_{str} \qquad (5.1)$$

And since salience reflects local differences between geographic objects, i.e., one object being noticeably different in one or several of the salience properties, the final step is to calculate these differences, for example, by employing maximum or minimum operations, which will result in the geographic object with the highest (or lowest) salience value in a given local configuration of objects.

Let's look at an example (from Raubal and Winter [35]) to see how this formal model may be applied to determining landmark salience. Consider the situation depicted in Fig. 5.2.

There are three buildings that are of interest at the marked intersection: Café Aida, Bank Austria and the Haas building. Table 5.2 shows the values of each property for the Haas building. The table also indicates which of the properties are significant and, thus, need to be taken into account for calculating landmark salience. Calculation is done using Eq. (5.1). For the other two buildings, determining their landmark salience works accordingly, which leaves us with a salience score of 1.8 for the Haas building, 1.2 for Bank Austria, and 0.9 for Café Aida. Applying a maximum operation results in the Haas building as being the most salient geographic object at the intersection. It seems to be the most recognizable one in terms of landmark identification. We will see below that this does not automatically mean it is always the best landmark to use when communicating about this intersection.

Raubal and Winter's model [35] has seen several extensions over the years. Emphasizing the aspect of structural attraction, *advance visibility* [52] is one such extension. The underlying assumption of advance visibility is that landmarks that

Table 5.2 Deriving landmark salience for the 'Haas' building

Measure	Property	Value	Significance	Attractiveness Measure	Weight	Total
Visual Attraction	p_{vf}	17400	1	$s_{vis} = 4/5$	$w_{vis} = 1$	1.8
	p_{vsf}	0.62	1			
	p_{vsd}	0	0			
	p_{vc}	[21,24,38]	1			
	p_{vv}	10600	1			
Semantic Attraction	p_{sec}	1	1	$s_{sem} = 2/2$	$w_{sem} = 1$	
	p_{sem}	1	1			
Structural Attraction	p_{stn}	–	0	$s_{str} = 0/2$	$w_{str} = 1$	
	p_{stb}	–	0			

Example taken from [35]

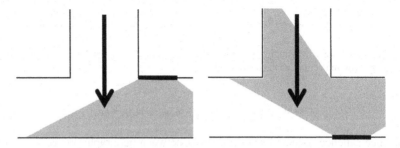

Fig. 5.3 A landmark with low (*left*) and high (*right*) advance visibility; modified from [52]

are identifiable early on along a route are more useful than those that can only be spotted at the very last moment. Figure 5.3 shows an example of a landmark with low and high advance visibility.

Formally, advance visibility v is a measure of *route coverage c* and *orientation o* of the landmark in question. Route coverage reflects how much of a landmark is visible along the entering route segment when approaching the intersection where it is located. The route coverage measure is calculated as the ratio between the part covered by the landmark and the total length of the route segment, which is specified by start and end point p_s, p_e. Orientation o measures the orientation of a landmark towards the route. A landmark that is oriented in moving direction is easier to spot than one that requires considerable head movement. Formally, orientation is defined as the difference between façade orientation d_f and route segment orientation d_r, both in terms of cardinal directions.

$$v = c * o \qquad (5.2)$$

$$c = \frac{|p_i p_e|}{|p_s p_e|}$$

$$o = \frac{|d_f - d_r|}{180}$$

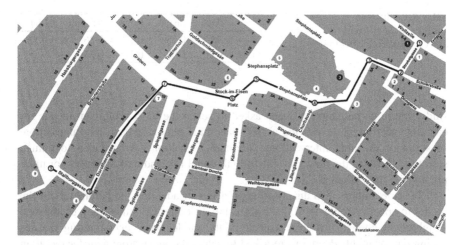

Fig. 5.4 The route investigated in [31]. *White dots* indicate differing selections between participants and model, *black dots* coinciding selection. Modified from [31]

Let us return to the example of Fig. 5.2. If also accounting for advance visibility, it turns out that for the given direction of travel, the Bank Austria building has a higher salience value than the Haas building. Coverage for Bank Austria is $c = 1$, and the orientation is also close to 1, while the Haas building is orientated nearly completely away from the route. In principle, advance visibility needs to be calculated for every landmark from every possible direction of approach to cover for every possible wayfinding situation.

Klippel and Winter [26] presented a further extension of the salience model that expands the aspect of structural salience. It accounts for the location of landmarks along the route and the kind of wayfinding action that needs to be performed. Since this is route-specific and not a general property of a landmark candidate, their approach will be further discussed in Sect. 5.3.

Raubal and Winter's formal model has been empirically evaluated by Nothegger, Winter, and Raubal [31] (see also [53]). They implemented (partially manual) measures for façade area and shape, color, visibility, and semantic attraction of a building. Data for one route through Vienna's first district has been collected, combining several data sources (again, a partially manual process); see Fig. 5.4.

Using this data, for each intersection along the route the most salient landmark was calculated. These were then compared with the results of a human subject

Table 5.3 Different weightings for different aspects of salience, after [53]

	Area	Shape	Color	Visibility	Marks
Day	0.11	0.15	0.37	0.26	0.12
Night	0.26	0.00	0.21	0.23	0.30

study, where participants had been asked to select the most prominent façade for each intersection while viewing a 360° panorama photograph of that intersection. In seven of the nine intersections, automatic selection corresponded to human selection, showing the power of this formal model for landmark salience—or more precisely façade salience.

The same authors also discussed approaches of adapting the parameters of the model to specific contexts [53]. Again using an empirical study, they established weighting factors for different aspects of the model when encountering façades during day or night. People were asked to select the most prominent façade at an intersection from a photograph (showing either a day or a night shot), and then to rank different aspects according to their importance for making this selection. Table 5.3 lists these factors. While visibility is important both during the day and at night, shape does not seem to be used at all at night. Instead a façade's area becomes much more important. The same holds for marks, especially if they are illuminated.

Elias [11] identified this challenge of data collection and parameter adaptation as the weak spot of Raubal and Winter's salience model. She proposed to use existing topographic or cadastral data sets and to run machine learning approaches to identify potential landmark candidates. The attributes in these data sets can be used as feature vectors describing the different geographic objects. These feature vectors may be fed into classification or clustering algorithms, which will then identify 'outliers', i.e., geographic objects that are not easily joined with other objects. Arguably, these outliers stand out from their surrounding environment and, thus, can be considered to be landmarks following Presson and Montello's definition [32].

Depending on the chosen classification algorithm, using this approach may require to normalize the data first. Attributes may need to be preprocessed such that they are all on the same scale and use the same measurement type (ordinal, nominal, etc.). This is to ensure that no single attribute dominates the discrimination between objects. Elias focused on buildings as landmark candidates and proposed the following attributes to use in the classification (see Table 5.4). As can be seen, these attributes either refer to land use or are derived from the geometry of buildings, in other words information that can be expected from a cadastral data set. This makes data collection easier, however, it also means that visual attractiveness can at best only be implicitly inferred from non-visual attributes.

From the many possible approaches, Elias chose to test her approach using ID3 [33], a supervised classification algorithm, and Cobweb [13], a hierarchical clustering algorithm. While both approaches seem promising, ID3 has the advantage

Table 5.4 Attributes of buildings used in the classification of a landmark candidate (from [11])

Attribute	Description
Building use	Public, residential, Outbuilding, ...
Building label	Name or function of building
Size of building	Length * width in m^2
Elongation	Ratio length/width
Number of corners	Counting corners (normally 4 to 6)
Single building	All alone, single in a row, one Neighbor, ...
Building moved away from road	Closest distance to road in m
Ratio of building area to parcel area	$\dfrac{\text{Building ground area}}{\text{parcel area}}$
Density of buildings (direct neighborhood)	$\dfrac{\text{Number of buildings}}{100\,m*100\,m}$ in $\frac{1}{m^2}$
Density of buildings (district)	$\dfrac{\text{Number of Buildings}}{500\,m*500\,m}$ in $\frac{1}{m^2}$
Orientation to road	Along (length towards road), across (width), angular, building at corner (in grad)
Orientation to north	Angle building length to north in rad
Orientation to neighbor	Difference angle to neighbor in rad
Perpendicular angle in building	Deviation of angles to normal in rad
Parcel land use	Public, residential, commerce and service, industrial, ...
Number of buildings on parcel	Counting buildings
Special building objects on parcel	Number of car ports, winter gardens, ...
Neighbor land parcel use	0 or 1 (Boolean)
Form of parcel area	Number of corners, number of neighbors

of not only identifying salient buildings, but also making attributes contributing to a building's salience explicit. Thus, it would be easier to generate guidelines for landmark identification from ID3 than Cobweb.

The approaches discussed so far determine local landmarks, i.e., they allow identifying geographic objects that stand out from their immediate surroundings. However, they do not establish relationships between these landmarks and they do not rank which landmark is the dominating one for a given area, i.e., which best represents this area in any references made to it. Linking the conceptual ideas of Raubal and Winter's formal model [35] with Elias' data mining approach to determining local landmarks [11], Winter et al. [55] addressed exactly this problem. Their approach allows for generating a leveled hierarchy of local landmarks. Algorithm 5.1 illustrates their approach. Fundamentally, it is based on partitioning an environment using Voronoi diagrams [1] (Sect. 4.2.5). On the lowest level, the most salient landmark in each local neighborhood—say an intersection—is identified using an unsupervised adaptation of the ID3 algorithm (Line 1). These most salient landmarks form the next higher level in the hierarchy. Step 1 of the algorithm gets repeated until there is only one landmark left, i.e., the Voronoi diagram only consists of a single cell.

Algorithm 5.1: Algorithm to generate a leveled landmark hierarchy, from [55].

Data: l_k^i is the landmark k on hierarchy level i; L^i is the set of all landmarks at level i; L_k^i is the set of all landmarks in the immediate neighborhood of landmark l_k on level i.

1 **forall the** *landmark* l_k^i *at level* i, $l_k^i \in L^i$ **do**
2   Compute the most salient landmark $l_{k.max}$ in its immediate neighborhood
      $L_k^i = \{l_j \mid \text{dist}(l_j - l_k) \leq 1, l_j \in L^i\}$.
3   Add the most salient landmarks at level i as the set of landmarks at level $i + 1$:
   $\{l_{k.max} \mid l_k^i \in L^i\} \leftarrow \{l_r^{i+1}\}$.
4 **if** $|L^{i+1}| > 1|$ **then**
5   Compute the Voronoi partition and the Delaunay triangulation of L^{i+1} and go back to
      step 1.
6 **else**
7   Stop.

The emerging cells can be considered to be the *influence region* of a landmark [55]. As with most models, this strictly leveled approach to a landmark hierarchy simplifies reality. It is well conceivable to have more subtle variations in (perceived) landmark salience, which may lead to less strict hierarchies, or sub-hierarchies within a single level.

5.2.2   Data Mining Salience

So far, we have discussed methods and techniques to identify landmark candidates from geographic data. It is also conceivable to identify landmarks in less structured geographic content, for example, from images or texts that depict or describe spatial situations. In this case, methods from information retrieval, and geographic information retrieval [20] more specifically, come into play. These usually encompass some machine learning or clustering approaches to mine relevant information—here landmark candidates—from unstructured or semi-structured sources.

Such approaches aim at finding relevant information in the selected data sources where relevance depends on the question asked. For example, Zhang et al. [56,57] extracted route directions, in particular their destinations, from web sites. They achieved this by classifying elements of the text based on both the structure of the text and the HTML tags, similarity of elements to prototypical patterns (e.g., 'turn left/right at LOCATION'), the position of textual elements in the overall text, and other parameters. Similarly, Drymonas and Pfoser [7] reconstruct tourist walks from travel guides by extracting motion activities ('walk along...', 'turn left...') together with the associated points of interest. These points of interest in turn can be geocoded again, which allows for mapping the walks onto geographic data. While none of these approaches searches for landmark candidates explicitly, some of the methods would be applicable for such an aim as well.

Tezuka and Tanaka [47] explicitly aim at extracting landmark information from texts taken from the World Wide Web. They compared several statistical and linguistic measures to calculate the landmarkness of some geographic object mentioned in one or several documents of a given corpus. These measures are:

1. *Document frequency*: In how many documents does a term (reference to an object) appear;
2. *Regional co-occurrence summation*: How often does an object appear with its surrounding objects in a document;
3. *Regional co-occurrence variation*: With how many different objects does an object co-occur in documents;
4. *Spatial sentence frequency*: Using spatial trigger phrases, how often is an object used in a spatial sentence;
5. *Case frequency*: What grammatical structure is used to refer to an object.

Their results show that the measures with spatial context (2–4) match better with a human-judged set of landmarks than the one without spatial context (1,5). If only interested in the most prominent landmarks, regional co-occurrence summation is a useful measure, as it has high precision (the ratio of correctly retrieved objects to all retrieved objects) for low recall (the ratio of correctly retrieved objects to all existing correct objects). If a large set of landmark candidates is desired, sentence frequency yields the best results with a relatively high precision for high recall situations, i.e., many landmark candidates are retrieved of which many are correct.

Getting more visual, several approaches exist that use tag-based descriptions, photographs, or a combination of both to determine points of interest or specific relevant regions. Mummidi and Krumm [30] used pushpins on Microsoft's Bing Maps to find POIs that are not already contained in the underlying database. Each pushpin has a known position (coordinate), and an associated title and textual description. Pushpins are clustered based on their latitude and longitude, using a hierarchical agglomerative clustering technique [10]. This clustering technique starts out with each pushpin as its own cluster, and then iteratively combines closest clusters until only one cluster is left. When combining two clusters, the position of the emerging cluster is the centroid of all contained pushpins. Figure 5.5 illustrates this idea further.

To figure out whether a given cluster actually describes a POI, the authors make use of n-grams of the pushpins' descriptions. An n-gram is a phrase with n words in it. In this case it is a subsequence of n consecutive words from each description. For example, in the description 'my favorite pizza place' all valid 2-grams (called bigrams) would be 'my favorite', 'favorite pizza', and 'pizza place' (but not 'my place', because these are not consecutive words in the description). The main measure to identify useful clusters the authors use is *term frequency inverse document frequency* (TFIDF). This measure compares the number of times a specific n-gram appears in a cluster ('term frequency') with the number of times the same n-gram appears in all clusters combined ('document frequency'). Dividing the

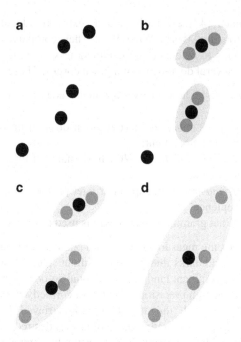

Fig. 5.5 A toy example for hierarchical clustering of POIs. Initially, every POI is its own cluster (**a**). In a second step, the closest two POIs form a new cluster. The new cluster center—the centroid—is shown in *black*, the original POIs in *dark grey* (**b**). This process is repeated until there is only a single cluster left (**c, d**)

former by the latter gives a number that indicates how unique a given n-gram is for a cluster because higher numbers indicate that this n-gram appears frequently within a cluster, but only sparsely outside it. In addition, Mummidi and Krumm also define a minimum size for a valid cluster, and measure 'term purity', i.e., the fraction of descriptions within a cluster that contain a given n-gram.

Others have used tags associated with Flickr photographs to find appropriate labels for places or to delineate city cores [18, 34], using similar ideas, but different techniques to those just presented.

Next, we will have a look at approaches that aim at identifying prototypical and/or prominent views for specific locations from photograph collections. For example, Kennedy and Naaman [23] as well as Zheng et al. [58] use unsupervised learning techniques to find canonical (prototypical) views for given clusters of photographs that (likely) show the same geographic object. They use Flickr[1] or Picasa[2] and Panoramio[3] as data sources, respectively, exploiting tags, geographic position, and the images themselves.

---

[1]http://www.flickr.com, last visited 8/1/2014

[2]https://picasaweb.google.com, last visited 8/1/2014

[3]http://www.panoramio.com, last visited 8/1/2014

Table 5.5 The five most prominent landmarks of the world and for some well known cities, according to the approach by [5]

	1st	2nd	3rd	4th	5th
World	Eiffel	Trafalgarsquare	Tatemodern	Bigben	Notredame
New York	Empirestatebuilding	Timessquare	Rockefeller	Grandcentralstation	Applestore
London	Trafalgarsquare	Tatemodern	Bigben	londoneye	Piccadilycircus
Rome	Colosseum	Vaticano	Pantheon	fontanaditrevi	Basilica
Berlin	Brandenburgertor	Reichstag	Potsdamerplatz	Berlinerdom	Tvtower

Photos taken with digital cameras may contain much more information than just the image itself. The Exchangeable image file format (Exif) standard specifies the format for images, as well as for sound and tags, used by digital cameras, including those in smart phones. This information encompasses the camera type and basic information about how the image was taken, namely, exposure, aperture and focal length, but also the date it was taken, the (GPS) position it was taken at (if the camera allows for this), the photo's resolution, the color space, the ISO setting, aspect ratio, the software used in the camera, and much more. In addition, today's photo websites store information about who has uploaded a photo (the user, either by name or ID), when it was uploaded, and which digital album it is contained in. Furthermore, these sites allow users to describe their photos by comments and/or tags. Tags are freeform keywords associated with an element—here a photo. Typically, users are free to choose whatever content they like in their tags.

Crandall et al. [5] may have presented the most advanced attempt to mine such digital photograph collections. Their approach can deal with millions of photos. The authors are able to identify the most representative images for the most prominent landmarks of the largest US or European cities, and to track photographers through a city, for example. This approach exploits the so called *scale of observation*, i.e., the fact that on different levels of scale different effects are observable. Observing on the scale of countries, cities will appear as clusters, when observing on the scale of a single city, points of interest and tourist attractions will emerge.

Using an estimate of the scale, latitude / longitude values of the photos' locations are treated as points in a plane and clustered by the mean shift technique. This technique yields peaks in the distribution of photos. The magnitude of these peaks represents the number of different photographers that took a photo at this location. Table 5.5 shows some results for different cities. The labels are automatically generated by picking the top tag within a cluster according to TFIDF.

Fig. 5.6 The canonical views for several of Europe's most photographed cities; from http://www. cs.cornell.edu/w8/~crandall/maps/map-europe.png (last visited 15/11/2013). Figure used with permission by the author

> The mean-shift technique is a statistical approach for estimating parameters of an underlying probability distribution. It is non-parametric, which is advantageous because no input parameters have to be guessed or arbitrarily set. More specifically, mean-shift provides the modes. The mode is the value that appears most often in a distribution.
>
> In the case of finding the most popular landmarks, the underlying probability distribution of where in a space people take photos is unobservable, i.e., it is impossible to find a functional description of that distribution. Still, mean-shift allows for finding the modes, which correspond to those places where a lot of people take a photo, i.e., places that are interesting or important.

For the identified landmarks, Crandall et al. look for canonical views, i.e., typical photographs. This combines clustering with graph construction. Images are tested for similarity using SIFT (Scale Invariant Feature Transform; see [28]) interest points. A graph is constructed with each photo as a node and each edge between a pair of nodes is weighted with the similarity between the two photos. The graph is partitioned to find clusters of similar photos. For each cluster the node (photo) with the largest weighted degree is chosen as the canonical view for this cluster. Figure 5.6 shows an example of canonical views for some of Europe's most popular cities in form of a map.

Obviously, all the approaches presented in this section can only find what is in the sources. Their results depend on the available data, i..e, on what people upload to the Internet. Thus, these approaches are very good at picking up the most prominent locations in a country or city, where there will be many uploads (often from tourists), but will fail to pick up local landmarks in some residential neighborhoods because only very few, if any, data is published about them. This issue will be further discussed at the end of this chapter in Sect. 5.5.

5.3 Landmark Integration

When a set of landmark candidates exists, these can be used in a range of services, some of which we will discuss in the next chapter. Most of the times, this will require selecting one—or a very few—landmarks from the set of all candidates. This is necessary because not all landmark candidates will be relevant for a given situation or task. And often it will be the case that the landmark that gets selected is not the most outstanding of the whole set, according to the salience measure applied. Rather, it will be the most *relevant* landmark. Accordingly, this section covers several approaches to determining relevance of a landmark candidate for specific tasks and situations, predominantly in navigation scenarios.

Klippel and Winter [26], for example, extended the salience model by Raubal and Winter [35] in order to select a landmark at an intersection that is most suitable to initiate a turning action (e.g., 'turn left' or 'veer right'). Figure 5.7 illustrates possible locations of landmarks with respect to turning at an intersection. Not all landmarks are equally suitable for identifying the required turn.

The location of a landmark relative to a turn at an intersection is an aspect of *structural* salience. The extended model uses advance visibility, as it has been discussed in the previous section, but additionally takes into account the

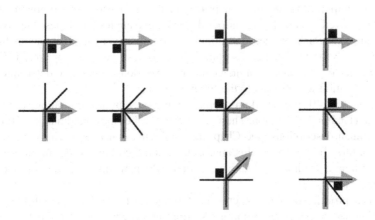

Fig. 5.7 Possible locations of landmarks with respect to turning actions; modified from [26]

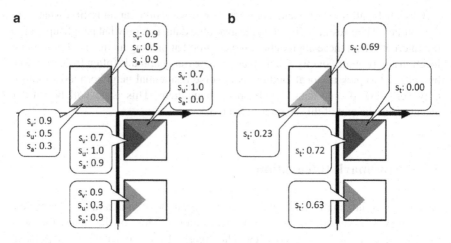

Fig. 5.8 (a) Turning right at an intersection with three buildings as landmark candidates (only relevant façades are considered); (b) the resulting overall salience values. Adapted from [26]

configuration of the street network and the route along this network. This results in a distribution of salience values for different landmark candidates, which allows for a selection of the most suited, i.e., most relevant landmark for a specific wayfinding situation. Usually, landmarks located at an intersection are structurally more salient than those between intersections along a street segment, and those passed before a turning action are more salient than those after a turn. Figure 5.8 illustrates a street intersection and the different salience values of the buildings located there. You can clearly observe the emergence of a salience distribution, with the façade facing the ingoing street segment of the building located before the intersection being the most salient one in this case.

This is an important insight. In wayfinding, relevance of a landmark depends on its location relative to the street network. Consequently, it is important to being able to compute this location. In principle, this is a geometric problem, which could be solved by calculating various distances and angles for all possible different configurations. Such geometric solutions are complicated by the fact that landmarks may not only be represented as points, but may also be linear or area-like [17]. This will have an influence on the geometric operations required to determine their relative location with respect to the street network.

Richter [36, 40] presented an alternative approach by using qualitative descriptions of a landmark's location. These descriptions reflect how landmarks are referred to in human descriptions (see Chap. 6), employing concepts, such as 'before' or 'after' a turn. In a nutshell, this approach exploits (circular) ordering information to determine a landmark's relative location, which works for different landmark geometries.

This approach—as many others—models a street network as a graph-like representation, where nodes represent intersections and edges represent street segments.

Fig. 5.9 Important concepts in modeling wayfinding in street networks: a decision point with two functionally relevant branches, the incoming and outgoing route-segment. From [36], modified

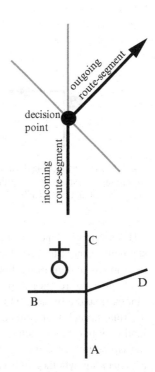

Fig. 5.10 Circular order of objects at an intersection. Starting with *A*, the order is *A* < *D* < *C* < church < *B* < *A* (from [36], modified)

Importantly, the representation is coordinate-based, i.e., each node has a coordinate, which then allows to talk about distances and directions. Considering movement through this network, an intersection becomes a decision point, i.e., a point along the way where alternative continuations of movement direction exist. Only the branches one enters and leaves this decision point on are *functionally* relevant. They are crucial in finding the (right) way. The former is called *incoming route segment*, the latter *outgoing route segment* (Fig. 5.9).

Landmarks are integrated into this representation by using the same reference frame as the network elements. They are represented either as point, (poly)line or polygon using the same coordinate system as for the decision points. This way, it becomes possible to relate landmarks to the route. In Richter's approach, to this end circular ordering information is exploited. Figure 5.10 illustrates this idea. Ordering information derives from the linear, planar, or spatial ordering of features [44]. This kind of information is a powerful structuring mechanism as it only requires knowledge about a neighborhood relation between relevant objects. No further metric knowledge is needed.

First, let us consider landmarks represented as points. Determining the location of linear and area-like landmarks works accordingly as will be explained below. There are three qualitatively different cases that need to be distinguished when looking at point landmarks. The turning action at the decision point may happen after the landmark has been passed, before it is passed, or the landmark may not be located at a functionally relevant branch at all (Fig. 5.11; see also [26]). In other words, the

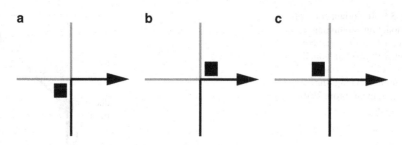

Fig. 5.11 Three functionally different locations of a point landmark relative to a decision point: (**a**) turning action after passing the landmark; (**b**) turning action before passing the landmark; (**c**) landmark not at a functionally relevant branch (from [36], modified)

landmark may be next to the incoming route segment, next to the outgoing route segment, or next to any of the other branches of the intersection. Accordingly, the task now is to determine this *next to* relationship.

A *next to* relationship corresponds to a neighborhood relation between a landmark and a branch, which gets us back to the circular ordering. By introducing a virtual branch that connects the landmark with the decision point we include the landmark into the branches' circular ordering. With that, we can determine whether the virtual branch is a neighbor to one of the functionally relevant branches, simply by checking whether one succeeds the other in the ordering. If the landmark is direct successor or predecessor of the incoming route segment, the turning action is performed after the landmark is passed (denoted with $lm^<$). Accordingly, if the landmark is a neighbor to the outgoing route segment, the turning action is performed before passing the landmark ($lm^>$). If the landmark is not directly neighbored to either of the functionally relevant route segments, all we can say is that a turning action is performed *at* the decision point with this particular landmark (denoted by lm^-).

One further restriction is illustrated in Fig. 5.12. A landmark may be next to the incoming route segment, but still only be passed after the turning action or not directly passed at all (e.g., at a T-intersection). Therefore, we need to further restrict the neighborhood region for incoming and outgoing route segment, such that only the area before or after the turning action, respectively, is considered. Again, this is solved by introducing virtual branches demarcating these regions. These virtual branches start at the decision point and are perpendicular to the incoming (or outgoing) route segment. We call the area next to the incoming route segment *before-region*, and the area next to the outgoing route segment *after-region* [40].

Using this combination of virtual branches and reasoning with ordering information we can determine where a point landmark is located relative to the route and, consequently, what its functional role is in terms of the wayfinding process (see Chap. 6). However, as stated above, landmarks are not always represented as points. They may also be extended objects, i.e., a line or polygon. Richter's approach can be extended to handle these objects as well in a straightforward manner [36].

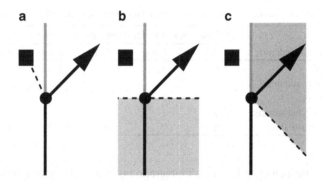

Fig. 5.12 Virtual branches demarcating before- and after-region are needed to correctly distinguish passing a landmark before or after the turning action. (**a**) An example case where a landmark is next to the incoming route segment, but actually not directly passed during the turning action; (**b**) the before-region for this situation; (**c**) the after-region

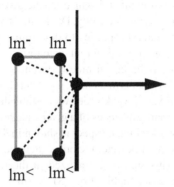

Fig. 5.13 Example of the relation of an extended landmark to the route. For each coordinate, its relation to the route is determined. Since different relations hold, the overall relation is lm^- (from [36], modified)

Given that these landmarks are represented by a sequence of coordinates, for each of these coordinates their relation to the decision point is determined (either $lm^<$, $lm^>$, or lm^-). If for every coordinate the relation is $lm^<$ ($lm^>$), the landmark is passed before (after) the turning action and we can assign the relation $lm^<$ ($lm^>$) to the landmark as a whole. In all other cases, i.e., if there is a mix between $lm^<$ and $lm^>$, or the relation is lm^- for some or all coordinates, then the landmark is not completely located in either the before- or after-region of the decision point. In this case, the relation lm^- should be assigned indicating that the actual location of the landmark is difficult to communicate precisely (and thus simply 'at' may be used to indicate the presence of the landmark; see Fig. 5.13).

Dale et al. [6] presented another approach integrating landmarks into route directions. Their CORAL system produces natural language instructions for route following, mimicking human principles of direction giving. It will be further dis-

Table 5.6 Landmark scoring system based on categories, after [9]

Typical landmark	Frequency of typical landmarks in category				
	All	Most	Many	Some	Few
Ideal	8	4	2	1	0
Highly suitable	4	4	2	1	0
Suitable	2	2	2	1	0
Somewhat suitable	1	1	1	1	0
Never suitable	0	0	0	0	0

cussed in the next chapter. Landmark integration is based on work by Williams [50]. The approach is not well documented, but seems to use the location of a landmark and the travel direction as parameters to determine whether a landmark is relevant for describing a given route.

As a final thought in this section, let us step away from looking at each individual geographic object in determining its relevance for the given context, and consider a simplified, more general approach. Recently, Duckham et al. [9] explored categories of features instead of individual properties to determine suitability as a landmark. They established a rank order of different categories, such as restaurants, gas stations, or schools, which is based on nine different aspects (e.g., physical size, proximity to road, visibility, or permanence). For each aspect default assumptions are made for every category. The rank order was established by a panel of experts, who also decided on how many instances of a given category are likely to be typical. The resulting scoring system for each aspect is shown in Table 5.6.

The overall suitability (or relevance) score is defined as the linear sum of all nine suitability aspects. This sum is normalized in the range [0, 1], with 1 being most suitable and 0 being least suitable (Eq. 5.3). Normalizing the weighting makes the score comparable between different settings and also independent from the absolute numerical values used in the expert rating. The score value $\text{score}_f(c)$ is the suitability score from Table 5.6 for a landmark category $c \in C$ with respect to the suitability aspect f. F is the set of all nine suitability aspects.

$$w(c) = \frac{\sum_{f \in F} \text{score}_f(c) - \min(\{\sum_{f \in F} \text{score}_f(c')|c' \in C\})}{\max(\{\sum_{f \in F} \text{score}_f(c')|c' \in C\})} \qquad (5.3)$$

Using categories instead of individuals has been implemented in the WhereIs route planner[4], a commercial route planning website in Australia owned by the same company that produces the yellow pages. Hence, these yellow pages are used as source for establishing suitable categories, and as data source for the locations of landmark candidates. The route planning algorithm calculates a shortest path using some standard shortest path algorithm (Dijkstra or A^*) and then finds all landmark

---

[4]http://www.whereis.com.au, last visited 8/1/2014

Fig. 5.14 Parameters
determining the weights in
the landmark spider approach.
A landmark's salience (s), as
well as its relation to the route
(orientation o and distance d)
are taken into account;
after [4]

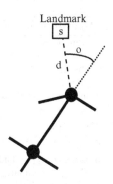

candidates (points of interest) along the route. At each intersection with at least one landmark candidate, the one with the highest weight gets selected—if two or more landmarks have the same weight, an arbitrary decision is made.

5.3.1 Landmark Integration in Routing Algorithms

Selection processes for landmark candidates have also been integrated into routing algorithms. These routing algorithms aim at a route that is easy to follow, as opposed to the shortest or fastest route that standard routing algorithms deliver. As previously discussed and further elaborated in the next chapter, (references to) landmarks make it easier for people to find their way.

These routing algorithms typically use the Dijkstra shortest path algorithm, but instead of simply using geometric distances, weights are based on cognitive criteria that determine the ease of finding the way. In the following, two such approaches are presented in some more detail: the landmark spider by Caduff and Timpf [4] and simplest instructions by Richter and Duckham [39].

Caduff and Timpf's algorithm aims at guiding wayfinders past landmarks at decision points. For each decision point, it determines the most relevant landmark, using its relevance as weighting parameter. More formally, the weight of an edge is the weighted sum of a landmark's distance to a decision point, the orientation of a traveler with respect to the landmark, and the landmark's salience (Eq. 5.4).

$$w_i = a * \text{Distance} + b * \text{Orientation} + c * \text{Salience} \tag{5.4}$$

The general idea of this weighting function is illustrated in Fig. 5.14. Closer landmarks are taken to be more relevant than landmarks farer away from a decision point [49]. Similar to the concept of advance visibility, landmarks in movement direction are easier to spot, and thus more relevant than those located behind a traveler.

However. since the Dijkstra algorithm uses a minimum-weight approach, weights have to be converted such that highly relevant landmarks get a very low weight,

---

Algorithm 5.2: The landmark spider routing algorithm.

Data: $G = (V, E)$: a directed graph, $s \in V$: the start vertex; ε: the set of outgoing edges over all nodes; $w : \varepsilon \to \mathscr{R}^+$: the weighting function.

Result: $c_s : E \to \mathscr{R}^+$: the minimal cost (clearest path) to all nodes from origin s.

1 S={} ;   /* *The set of visited edges.* */
2 **forall the** $e \in E$ **do**
3 $\quad \lfloor \quad c_s(e) \leftarrow \infty$;   /* *Initialize the costs to reach an edge for all edges.* */
4 **forall the** $(s, v_i) \in E$ **do**
5 $\quad \lfloor \quad c_s((s, v_i)) \leftarrow w((s, v_i))$;   /* *Initialize outgoing edges of s with their weight.* */
6 **while** $|E \backslash S| > 0$ **do**
7 $\quad e_{min} \leftarrow \min(c_s(e)|e \in E \backslash S)$;   /* *Find edge with minimal costs* */
8 $\quad S \leftarrow S \bigcup \{e_{min}\}$;   /* *and visit it.* */
9 $\quad$ **forall the** $e' \in E \backslash S$ **do**
10 $\quad\quad$ **if** $(e, e') \in \varepsilon$ **then**
11 $\quad\quad\quad \lfloor \quad c_s(e') \leftarrow \min(c_s(e'), c_s(e_{min}) + w(e'))$; /* *Update the costs to all non-visited neighbors e' of e.* */

---

while less relevant landmarks receive a higher weighting. That is, weights here serve as a penalty; good landmarks mean low penalty. The according algorithm is listed in Algorithm 5.2.

Richter and Duckham's [39] algorithm is more complex, but essentially does something similar. The algorithm is based on Richter's work on context-specific route directions [37] (see next Chapter), which not only integrate landmarks into the instructions, but also minimize the number of instructions. The overall aim is to generate instructions that are memorizable and easy to follow. This is reflected in the algorithm.

This algorithm operates on the *complete line graph* [8, 51]. The complete line graph reflects movement options in a graph, i.e., it captures all possibilities to turn from one edge onto another. Formally, this is the graph $G' = (E', \varepsilon)$. E' is the set of edges in G, where the direction of edges is ignored (i.e., $(v_i, v_j) = (v_j, v_i)$ in E'). ε is the set of pairs of vertices in E that share their middle vertex, i.e., $\varepsilon = \{((v_i, v_j), (v_j, v_k)) \in E \times E\}$. A pair of adjacent edges in E represents a decision an agent can take to move from one edge to the next. Figure 5.15 illustrates these definitions further.

Richter and Duckham's algorithm models the instructions required to describe a route as a set I of instruction labels. Instructions are associated with decisions (the pairs of adjacent edges). These instructions describe what to do in order to move from one edge to another (e.g., 'turn right at this intersection'). Specifically these instructions may contain references to landmarks ('turn right at the church'). There may be more than one instruction associated with a pair of edges, reflecting alternative ways of describing the decision. And, likewise, the same instruction may be associated with several pairs of adjacent edges—there may be many intersections in the network where a right turn is possible. However, the algorithm assumes that

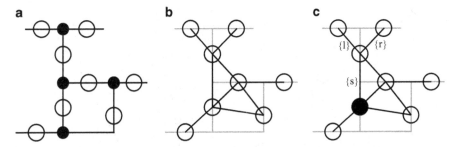

Fig. 5.15 The complete line graph and how it applies to the labeling and decision function in the simplest instruction algorithm. (**a**) An example street network; the *solid nodes* represent the intersections (vertices), the *hollow nodes* are the vertices of the line graph (the edges in the original graph). (**b**) The actual complete linegraph; the original graph is shown in *light grey* for reference. (**c**) Instruction labeling for some of the edges as seen from the solid vertex. Applying the decision function d(*solid node, s*) when at this solid vertex would result in the vertex just above it

there is no ambiguity in instructions, i.e., 'turn right' may not describe two or more decisions from the same edge.

Formally, instructions are covered by the labeling function $l : \varepsilon \rightarrow I^2$ and the decision function $d : E \times I \rightarrow E \cup \{\emptyset\}$. For a given edge e and an instruction i, $d(e, i) = e'$ gives the edge e' that results from executing the decision i at e, which may be the empty set indicating that instruction i is not possible to execute at e. Each instruction has a cost associated with it, represented by the weighting function $w : I \rightarrow \mathscr{R}^+$. Costs reflect the cognitive effort to execute an instruction, i.e., how difficult it is to understand and to identify what to do in the actual environment. Figure 5.15 shows an example that illustrates these concepts.

As a further feature of this algorithm, which makes it complex in the end, instructions may be 'spread forward through the graph'. This reflects the fact that often a single instruction may cover several intersections, such as in 'turn left at the third intersection' where implicitly the instruction tells a wayfinder to go straight at intersection one and two [25]. The cognitive and communication benefits of combining instructions this way will be further discussed in Chap. 6. Here, it is only important to note that there are limits to combining instructions (e.g., nobody would instruct someone to 'turn left at the 47th intersection'). Therefore, in spreading instructions through the graph, a distinction is needed between an edge being *reachable* and being *chunkable*. An edge e_t is *reachable* from an edge e_s with an instruction i if there exists a path from e_s to e_t that can be encoded as a sequence of executions of instruction i. The same edge is *chunkable* from e_s if the sequence of instruction i required to reach e_t from e_s is valid according to the combination rules. This is checked by the validity function $v : E \times E \rightarrow \{true, false\}$.

An instruction gets spread forward as long as there are edges reachable with it. A cost update is then only performed for those edges that are chunkable. This way

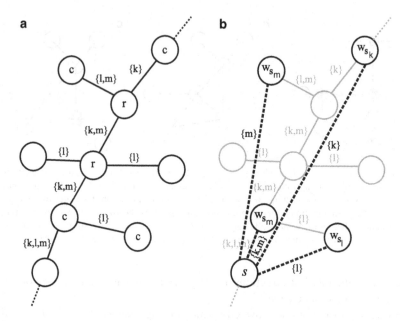

Fig. 5.16 Spreading instructions forward in a graph; edges are labeled with their associated instruction sets. (**a**) Reachable vertices (r) from s as here the instructions match. Chunkable vertices are denoted by c; here instructions adhere to the combination rules. (**b**) Chunkable edges can be reached in a single step, which corresponds to dynamically introducing new edges between the first vertex and the chunkable vertex. Costs for reaching these edges are equivalent to reaching the first edge (denoted by the different w_s)

of spreading instructions can be viewed as dynamically introducing new edges to the graph that connect first and last vertex of a valid combination of vertices (see Fig. 5.16).

The algorithm to determine simplest instruction paths is presented in Algorithm 5.3. The algorithm generates a predecessor function $p : E \to E \times I$ that stores for each edge the preceding edge in the least cost path and the instruction used to reach this edge from its predecessor. If several edges are chunkable by one instruction, the predecessor is the first edge of the combined edges (see Fig. 5.16). Accordingly, whenever an edge e is visited, for every instruction holding for (e, e') it is checked whether it lowers the costs associated with e' and whether this instruction can be used to reach other edges from e' as well (lines 11–12). To keep track of which instructions need to be further considered at an edge, all instructions the edge has been reached on so far are stored in a set U_e of instruction/edge pairs (i, e) (initialized in line 3). Next, the algorithm updates the weights associated with the unvisited edges that are incident with the current edge e (lines 13–17, as in the classic Dijkstra algorithm). Finally, the spreading of instructions forward through the graph is performed (lines 18–29). All edges that are reachable store information about the instruction being spread (line 25), but only those for which the resulting instruction would be valid according to v have their weights updated (lines 27–28).

Algorithm 5.3: Simplest instruction algorithm after [39].

Data: $G = (V, E)$: a connected, simple, directed graph; $G' = (E', \mathscr{E})$: the complete line graph of G; $o \in E$: the origin (starting) edge; I: a set of instructions; $w : I \to \mathbb{R}^+$: the instruction weighting function; $l : \mathscr{E} \to I^2$: the labeling function; $d : E \times I \to E \cup \{\varnothing\}$: the decision function; $v : E \times E \times I \to \{\text{true}, \text{false}\}$: the chunk validity function.

Result: Function $p : E \to E \times I$ that stores for each edge the preceding edge and the instruction used in the least cost path.

```
 1  forall the e ∈ E do
 2  │   c(e) ← ∞;                                          /* Initialize cost function c : E → ℛ⁺. */
 3  └   Uₑ ← ∅;                                            /* Set that stores instruction, edge pairs. */

 4  S ← {};                                                          /* Set of visited edges. */
 5  c(o) = 0;                                                        /* The cost to reach o is 0. */
 6  p(o) ← (o, i);                          /* o does not have a predecessor; i is arbitrary. */
    // Process lowest cost edge until all edges are visited
 7  while |E \ S| > 0 do
 8  │   e ← min(c(e)|e ∈ E \ S);                            /* Find edge with minimal costs. */
 9  │   S ← S ∪ {e};                                        /* Add e to visited edges. */
10  │   forall the e' ∈ E \ S such that (e, e') ∈ ℰ do
    │   │   // Update instruction/edge pairs from e to e'
11  │   │   forall the i ∈ l(e, e') do
12  │   │   └   Uₑ' ← Uₑ' ∪ {(i, e)};

13  │   forall the e' ∈ E \ S such that (e, e') ∈ ℰ do
    │   │   // Update costs to e' based on instruction weights
14  │   │   forall the i ∈ I such that d(i, e) = e' do
15  │   │   │   if c(e') > c(e) + w(i) then
16  │   │   │   │   c(e') ← c(e) + w(i);
17  │   │   │   └   p(e') ← (e, i);

    │   // Propagate instructions forward through the graph.
18  │   forall the (i, eₚ) ∈ Uₑ do
19  │   │   eₓ ← e';                                        /* Start propagating at e'. */
20  │   │   X ← S ∪ {∅};                       /* X is the set of all visited edges during propagation. */
21  │   │   while eₓ ∉ X do
22  │   │   │   X ← X ∪ {eₓ};                               /* Add eₓ to the visited edges. */
23  │   │   │   Set eₙ ← d(eₓ, i); /* eₙ is the next edge from eₓ reachable by instruction i. */
24  │   │   │   if eₙ ≠ ∅ then
25  │   │   │   │   Uₑₙ ← Uₑₙ ∪ {(i, eₚ)};  /* Add (i, eₚ) to the set of instructions of how to
    │   │   │   │   reach eₙ. */
26  │   │   │   │   if c(eₙ) > c(e') ∧ v(eₚ, eₙ, i) = true then
27  │   │   │   │   │   c(eₙ) ← c(e');                       /* Update the costs to reach cₙ. */
28  │   │   │   │   └   p(eₙ) ← (eₚ, i);

29  │   │   └   eₓ ← eₙ;                                    /* Continue propagation from eₙ. */
```

Algorithm 5.4: Algorithm for reconstructing the simplest instruction path, after [39].

Data: $G = (V, E)$: a connected, simple, directed graph; $o \in E$: the origin edge; $t \in E$: the target (destination) edge; $d : E \times I \rightarrow E \cup \{\emptyset\}$: the decision function; $p : E \rightarrow E \times I$: the predecessor function (generated by Algorithm 5.3).

Result: A sequence (word) P of edges corresponding to the optimal path and a sequence (word) L of labels corresponding to the best sequence of instructions.

```
1   P ← λ ;                                          /* Initialize the variables. */
2   L ← λ;
3   T ← λ;
4   e ← t;                                      /* Set e to the target (start from the back). */
5   while e ≠ o do
6   |   p(e) ← (e_p, i);                              /* Retrieve the predecessor of e */
7   |   L ← i + L;                          /* Add to L the instruction to get from e_p to e. */
8   |   e' ← e_p;
9   |   T ← e';                               /* The sequence of all edges between e_p and e. */
10  |   while (e', e) ∉ E do
11  |   |   e'' ← d(e', i);                      /* The edge reachable from e' via label i. */
12  |   |   T ← T + e'';                             /* e'' is between e' and e. */
13  |   |   e' ← e'';
14  |   P ← T + P;                                   /* Add T to the path. */
15  |   e ← e_p;                              /* Continue reconstruction from e_p. */
```

Algorithms 5.2 and 5.3 establish the costs to reach specific edges (or vertices) in the underlying graph. They do not actually return a path to this destination edge. This requires another reconstruction algorithm, which starts from the destination and walks backwards through the graph until it reaches the origin, assembling the path along the way. Algorithm 5.4 exemplifies such an approach for the simplest instructions path. Using an algebraic language notation, where E and I form alphabets, the algorithm constructs a path by iteratively prepending letters to an initially empty word. Retrieval of the instructions that need to be followed simply requires direct backtracking through the predecessor list (line 7), while reconstructing the edges to follow requires an additional loop to retrieve those edges between subsumed instructions (lines 8–15).

5.4   A Comparison of Landmark Identification and Landmark Integration Approaches

Richter [38], attempting to identify why landmarks have not been taken up in commercial navigation services so far, categorized several of the approaches presented in Sects. 5.2 and 5.3 with respect to different aspects. This categorization is summarized in Table 5.7. Most of the data mining approaches of Sect. 5.2.2 are not discussed here since they do not explicitly target landmarks of the scale and kind addressed in this book.

Table 5.7 Matrix of approaches to landmark identification and integration; modified from [38]

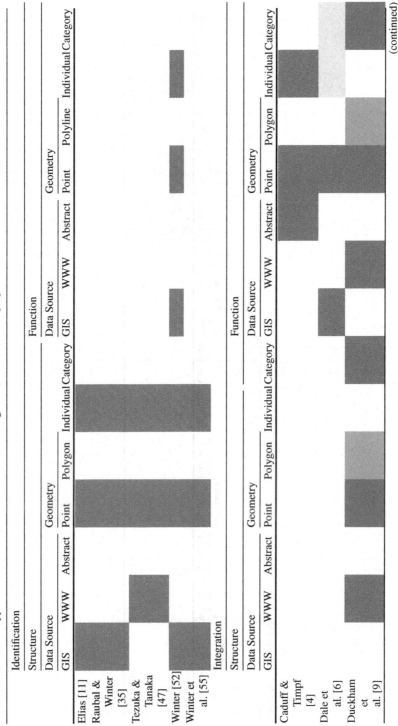

(continued)

Table 5.7 (continued)

	Identification							Function						
	Structure							Structure						
	Data Source		Geometry			Individual	Category	Data Source		Geometry			Individual	Category
	GIS	WWW	Abstract	Point	Polygon			GIS	WWW	Abstract	Point	Polyline		
Klippel & Winter [26]										■	■		■	■
Richter [36, 40]										■		■	■	■

A first broad distinction is made between approaches to landmark identification and to landmark integration. Some approaches focus on properties of the landmark candidates—how do they differ from other geographic objects in their surrounding? This corresponds to a static view on landmarks, similar to what Klippel describes as *structure* in wayfinding [24]. Other approaches account for the location of landmark candidates along a route to assess their suitability, which corresponds to a dynamic view or *function* in Klippel's terms. Further, approaches differ in the (assumed) source of data they use. Some approaches use 'classical' spatial databases of the kind attached to a typical GIS. Some use some web harvesting technique to source data from the web, either from general websites or from web catalogs, such as the yellow pages. Others do not specify where their data comes from, which has been marked as *abstract* in Table 5.7. Approaches also may differ in the geometry that landmark candidates can have. A majority assumes them to be point-like, however, some also consider more complex geometries (e.g., polygons). Finally, approaches may aim to identify individual objects (*instances*) or categories of objects (*types*).

If we look at Table 5.7 the first thing to observe is that approaches to landmark identification predominantly use a static view on landmarks while landmark integration adopts a dynamic view. This makes sense since landmark identification needs to find all geographic objects that may serve as a landmark in principle. These approaches need to assess the salience of objects, i.e., check whether an object sufficiently stands out from its local surroundings. As long as objects do not change, this has to be done only once and can be run as a preprocessing step in an actual system.

Landmark integration looks for landmarks that are actually useful in a given context. The selected landmarks need to be visible, sensibly describe the given situation, and support conceptualizing what to do in the situation. These are functional characteristics, as they depend on the specific situation, such as the current route to follow. As discussed, landmark integration does not necessarily choose the most salient landmark, but the most relevant one. Establishing the context, i.e., specifying to a sufficient degree the parameters influencing the selection process, is a major challenge here (see also the final discussion in Chap. 7).

Consequently, landmark identification and landmark integration are often seen as independent steps (e.g., [12]). This view becomes apparent from Table 5.7 as well. Each approach for landmark identification uses a concrete data source. However, most approaches to landmark integration do not specify the data source used. These approaches typically assume a set of landmark candidates to be given, i.e., some kind of landmark identification has to have happened previously.

It can further be observed that all approaches except [9] use individuals rather than types when assessing an object's suitability as landmark. And almost all approaches follow Lynch's tradition of assuming landmarks to be point-like objects [29]. While Duckham et al. [9] acknowledged that other geometries for landmarks may be useful, only Richter [36, 40] fully implemented an approach to determine the role of landmarks with different geometries.

5.5 Criticisms of Existing Approaches

As has been shown in this chapter's discussion so far, existing approaches to landmark identification and landmark integration are not really integrated. You may consider this a minor issue because it seems that an integration is a fairly straight-forward engineering task. But such a lack of integration certainly has prevented widespread application of the presented approaches beyond basic research. There are other more serious challenges for the widespread use of landmarks in location-based services, though. These will be further discussed in this section.

5.5.1 Data Challenge

Commercial systems use efficient algorithms to calculate shortest or fastest paths based on metric distances and using references to street names, which are easily extractable from a geo-referenced network representation of the street layout. In Sect. 5.3, we discussed some algorithms that account for landmarks in path search. In order to make this efficient, these landmarks need to be embedded into the existing network structure, i.e., graphs need to be annotated with objects that may serve as landmark candidates. Some systems utilize points of interest, such as gas stations or hotels. These are a potential source for landmark candidates (as implemented in the WhereIs routing service), but mostly POIs serve as selectable destinations or commercial announcements. Accordingly, there is a bias towards specific categories of objects. Further, their distribution and density will vary greatly—there are more POIs in a city center compared to a suburb or some rural farm land. This uneven distribution has consequences for the quality of landmark-based services.

Integrating landmarks into the network structure used for path search requires suitable data structures. The Urban Knowledge Data Structure discussed in Chap. 4 is a candidate for such a structure. However, even with such a data structure at hand, there is still the need to fill it with actual data. For most of the presented approaches, this is a highly data intensive process. As most approaches are based on individuals, individual objects have to be described in great detail, which is especially true for the calculation of façade salience as discussed in Sect. 5.2. The required information is hard to collect automatically and, thus, labor intensive, which makes this an expensive endeavor. It is therefore unlikely that such a collection will ever materialize on a commercial scale and lead to a city-, nation-, or even world-wide database of landmark candidates due to the immense collection efforts and costs.

This data challenge has been further discussed by Sadeghian and Katardzic [43], and Richter [38].

5.5.2 Taxonomies

Given the previous discussion, using types rather than individuals seems to be the more promising approach. Here, properties of individual geographic objects do not matter as they are inferred by some heuristics. These heuristics make use of a general assessment of a specific type's suitability as a landmark. For example, it may be argued that in general a pub is more salient than a doctoral clinic. Consequently, much less data would be required to make such an approach work. In fact, a POI database that categorizes its entries according to type would suffice. This is the approach taken by Duckham et al. [9], which has been discussed earlier in this chapter.

Götze and Boye [16] recently suggested using a machine learning approach to determine suitable landmark candidates. In their approach, landmark candidates are described using a feature vector that may contain data, such as distance from the route, but also categorical information, for example, whether the candidate is a restaurant. Their approach determines user preferences for landmarks from route directions generated by the users themselves. Depending on which candidates appear in these descriptions, preferences for specific kinds of landmarks are learned by the system and offered in system-generated descriptions in the future. This takes away the necessity for a detailed data collection for every landmark candidate since the feature vector uses fairly simple attributes. Still, these vectors need to be filled and candidates need to be collected for the approach to work—and most importantly, users need to be motivated to describe routes to themselves.

Even more, as argued above, existing POI databases usually exhibit a sparse and uneven distribution of POIs across the space they cover. For example, Richter [38] has shown for the WhereIs routing service that assuming an even distribution across Australia—which is clearly not the case—there would only be one POI object every 45 km^2. Thus, clearly new ways of collecting a sufficient number of landmark candidates with sufficient detail are required. These will be discussed in the final section of this chapter.

5.5.3 Crowdsourcing as an Alternative Approach

Crowdsourcing [46] is an approach to acquire desired content or services from a large number of people ('the crowd') rather than from traditional suppliers ('the individual'). It is a predominant approach in open source software development, where a (large) number of people interact and co-develop a software project.

When it comes to collecting potential landmark candidates we may use similar approaches as in crowdsourcing. Since our interest is rather in data (or information) than services, we may rather speak of *user-generated content* [27] or *volunteered geographic information* [15]. These three terms are not synonymous, however for the purposes of the following discussion the differences between them are not important and we will simply stick with user-generated content. Essentially, the idea is to access users' knowledge about an environment in order to collect and update the required information; the users serve as 'database' [41].

Holone et al. [19], for example, suggested a system that allows to mark route segments as bad or inaccessible for wheelchair users—or people with similar movement restrictions, even if only temporarily (e.g., pushing a baby stroller). Karimi et al. [21,22] discussed *SoNavNet*, a social navigation network, where people can provide and request recommendations for POIs and routes to these POIs. While neither of these approaches specifically targets landmarks, both exploit users and their willingness to contribute in order to provide better navigation services.

One way to motivate users to contribute the kind of data a service provider is looking for is to set up an entertaining incentive, such as a mobile game, in a way that the sought for data becomes a by-product of that game [2,54]. Bell et al. [2] designed such a game, called *EyeSpy*, which collects photographs of city locations that support navigation. In that game, players take photographs within a city environment and/or produce text tags describing the environment. These photographs are geo-referenced by using a phone's WiFi or GPS sensor readings. Other players then need to confirm these photographs by moving close to the location where they were taken and confirming that they have found what is depicted on the photographs. Players get points for performing such confirmations, but they also get points based on the popularity of a photograph, i.e., how often it gets confirmed by other players. The idea behind this setup is that photographs become more popular if they are easy to find and recognize. Therefore, in order to get more points, players will aim at submitting photographs they believe to be easily recognizable. Such photographs will also be easily recognizable in other contexts, for example, when provided as navigation assistance. Thus, the useful by-products of EyeSpy are photographs of a city environment that make specific locations more recognizable, i.e., have a landmark character.

Richter and Winter [14, 42] proposed mechanisms to collect landmark data in OpenStreetMap. OpenStreetMap (OSM) is a topographic data set of the world compiled from user-generated content. The OpenLandmarks application would allow users to mark existing objects within the OSM data as landmarks. A first prototype has been developed (Fig. 5.17) and used in some early experimentation. Currently, the system's interface is map-based. If users identify a building as a potential landmark while walking through a (city) environment, they may request all buildings in their surroundings that are actually represented as polygons in OSM to be highlighted as potential landmark candidates. Users may then select the one they have identified on the map and describe its landmark characteristics either by name (e.g., 'Flinders St Station') or by description ('big yellow train station').

It is envisioned to provide an improved version of such a tool to the Open-StreetMap community—or in fact anyone interested in contributing—and this way to build up a database of user-generated landmark candidates over time, similar to what has been achieved with OpenStreetMap. While we believe that employing mechanisms of user-generated content is the only realistic way of ever getting a

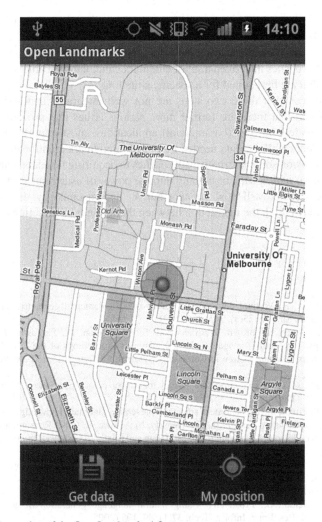

Fig. 5.17 Screenshot of the OpenLandmarks 1.0 prototype

sufficiently detailed and up-to-date data set of landmark candidates, there are many challenges attached to such an approach. These will be further discussed in the conclusions of this book in Chap. 7.

5.6 Summary

This chapter looked at computational approaches to handling landmarks. We pointed out that there are two steps required when planning to use landmarks in any kind of service: (1) the identification of potential landmark candidates; (2) the integration of the most relevant candidates into the offered service. We discussed several approaches that allow for identifying landmark candidates, either using geographic data or less structured data (texts, photographs) as a source. We then presented approaches that given a set of landmark candidates are able to select those candidates that are best suited for a specific situation.

We have seen that these two steps are not well integrated in today's existing approaches and that there are further more serious issues that prevent landmarks from being widely used in (commercial) applications, most importantly the huge effort of collecting all the data needed for a useful landmark identification. This led us to an outlook on some alternative approaches to acquiring this data, namely using types instead of individuals, or tapping into the power of user-generated content.

The next chapter will now connect human and machine by discussing how landmarks may enrich the interaction between them—in both directions.

References

1. Aurenhammer, F.: Voronoi diagrams: A survey of a fundamental geometric data structure. ACM Comput. Surveys **23**(3), 345–405 (1991)
2. Bell, M., Reeves, S., Brown, B., Sherwood, S., MacMillan, D., Ferguson, J., Chalmers, M.: Eyespy: Supporting navigation through play. In: Proceedings of the 27th International Conference on Human Factors in Computing Systems, CHI, pp. 123–132. ACM, New York (2009)
3. Buchsbaum, G.: A spatial processor model for object colour perception. J. Franklin Inst. **310**(1), 1–26 (1980)
4. Caduff, D., Timpf, S.: The landmark spider: Representing landmark knowledge for wayfinding tasks. In: Barkowsky, T., Freksa, C., Hegarty, M., Lowe, R. (eds.) Reasoning with Mental and External Diagrams: Computational Modeling and Spatial Assistance-Papers from the 2005 AAAI Spring Symposium, pp. 30–35. Menlo Park, CA (2005)
5. Crandall, D., Backstrom, L., Huttenlocher, D.P., Kleinberg, J.: Mapping the world's photos. In: Maarek, Y., Nejdl, W. (eds.) International World Wide Web Conference, pp. 761–770. ACM Press, Madrid, Spain (2009)
6. Dale, R., Geldof, S., Prost, J.P.: Using natural language generation in automatic route description. J. Res. Pract. Inform. Tech. **37**(1), 89–105 (2005)
7. Drymonas, E., Pfoser, D.: Geospatial route extraction from texts. In: Proceedings of the 1st ACM SIGSPATIAL International Workshop on Data Mining for Geoinformatics, pp. 29–37. ACM, New York (2010)
8. Duckham, M., Kulik, L.: "Simplest" paths: Automated route selection for navigation. In: Kuhn, W., Worboys, M., Timpf, S. (eds.) Spatial information theory. Lecture Notes in Computer Science, vol. 2825, pp. 169–185. Springer, Berlin (2003)
9. Duckham, M., Winter, S., Robinson, M.: Including landmarks in routing instructions. J. Location-Based Services **4**(1), 28–52 (2010)
10. Duda, R., Hart, P.: Pattern Classification and Scene Analysis. Wiley, New York (1973)
11. Elias, B.: Extracting landmarks with data mining methods. In: Kuhn, W., Worboys, M., Timpf, S. (eds.) Spatial information theory. Lecture Notes in Computer Science, vol. 2825, pp. 375–389. Springer, Berlin (2003)
12. Elias, B., Paelke, V., Kuhnt, S.: Concepts for the cartographic visualization of landmarks. In: Gartner, G. (ed.) Location Based Services & Telecartography-Proceedings of the Symposium 2005, Geowissenschaftliche Mitteilungen, pp. 1149–155. TU Vienna (2005)

13. Fisher, D.H.: Knowledge acquisition via incremental conceptual clustering. Mach. Learn. **2**(2), 139–172 (1987)
14. Ghasemi, M., Richter, K.F., Winter, S.: Landmarks in OSM. In: 5th Annual International OpenStreetMap Conference. Denver, CO (2011)
15. Goodchild, M.: Citizens as sensors: The world of volunteered geography. GeoJournal **69**(4), 211–221 (2007)
16. Götze, J., Boye, J.: Deriving salience models from human route directions. In: Proceedings of CoSLI-3 Workshop on Computational Models of Spatial Language Interpretation and Generation. Potsdam, Germany (2013)
17. Hansen, S., Richter, K.F., Klippel, A.: Landmarks in OpenLS - a data structure for cognitive ergonomic route directions. In: Raubal, M., Miller, H., Frank, A.U., Goodchild, M.F. (eds.) Geographic information science. Lecture Notes in Computer Science, vol. 4197, pp. 128–144. Springer, Berlin (2006)
18. Hollenstein, L., Purves, R.S.: Exploring place through user-generated content: Using Flickr to describe city cores. J. Spatial Inform. Sci. **1**(1), 21–48 (2010)
19. Holone, H., Misund, G., Holmstedt, H.: Users are doing it for themselves: Pedestrian navigation with user generated content. In: Proceedings of the 2007 International Conference on Next Generation Mobile Applications, Services and Technologies, IEEE Computer Society, pp. 91–99. Washington (2007)
20. Jones, C.B., Purves, R.S.: Geographical information retrieval. Int. J. Geogr. Inform. Sci. **22**(3), 219–228 (2008)
21. Karimi, H.A., Benner, J.G., Anwar, M.: A model for navigation experience sharing through social navigation networks (SoNavNets). In: International Workshop on Issues and Challenges in Social Computing (WICSOC2011). Las Vegas, NV (2011)
22. Karimi, H.A., Zimmerman, B., Ozcelik, A., Roongpiboonsopit, D.: SoNavNet: A framework for social navigation networks. In: Procceedings of the International Workshop on Location Based Social Networks (LBSN'09). ACM Press, New York (2009)
23. Kennedy, L.S., Naaman, M.: Generating diverse and representative image search results for landmarks. In: Proceedings of the 17th International Conference on World Wide Web, WWW, pp. 297–306. ACM, New York(2008)
24. Klippel, A.: Wayfinding choremes. In: Kuhn, W., Worboys, M., Timpf, S. (eds.) Spatial Information Theory. Lecture Notes in Computer Science, vol. 2825, pp. 320–334. Springer, Berlin (2003)
25. Klippel, A., Tappe, H., Habel, C.: Pictorial representations of routes: Chunking route segments during comprehension. In: Freksa, C., Brauer, W., Habel, C., Wender, K.F. (eds.) Spatial Cognition III. Lecture Notes in Artificial Intelligence, vol. 2685, pp. 11–33. Springer, Berlin (2003)
26. Klippel, A., Winter, S.: Structural salience of landmarks for route directions. In: A.G. Cohn, D.M. Mark (eds.) Spatial Information Theory. Lecture Notes in Computer Science, vol. 3693, pp. 347–362. Springer, Berlin (2005)
27. Krumm, J., Davies, N., Narayanaswami, C.: User-generated content. Pervasive Comput. **7**(4), 10–11 (2008)
28. Lowe, D.: Object recognition from local scale-invariant features. In: Proceedings of the Seventh IEEE International Conference on Computer Vision, vol. 2, pp. 1150–1157 (1999)
29. Lynch, K.: The Image of the City. The MIT Press, Cambridge, MA (1960)
30. Mummidi, L., Krumm, J.: Discovering points of interest from users' map annotations. GeoJournal **72**(3), 215–227 (2008)
31. Nothegger, C., Winter, S., Raubal, M.: Selection of salient features for route directions. Spatial Cognit. Comput. **4**(2), 113–136 (2004)
32. Presson, C.C., Montello, D.R.: Points of reference in spatial cognition: Stalking the elusive landmark. Br. J. Dev. Psychol. **6**, 378–381 (1988)
33. Quinlan, J.: Induction of decision trees. Mach. Learn. **1**(1), 81–106 (1986)
34. Rattenbury, T., Naaman, M.: Methods for extracting place semantics from Flickr tags. ACM Trans. Web **3**(1), 1:1–1:30 (2009)

35. Raubal, M., Winter, S.: Enriching wayfinding instructions with local landmarks. In: Egenhofer, M.J., Mark, D.M. (eds.) Geographic Information Science. Lecture Notes in Computer Science, vol. 2478, pp. 243–259. Springer, Berlin (2002)

36. Richter, K.F.: A uniform handling of different landmark types in route directions. In: Winter, S., Duckham, M., Kulik, L., Kuipers, B. (eds.) Spatial Information Theory. Lecture Notes in Computer Science, vol. 4736, pp. 373–389. Springer, Berlin (2007)

37. Richter, K.F.: Context-Specific Route Directions - Generation of Cognitively Motivated Wayfinding Instructions, vol. DisKi 314 / SFB/TR 8 Monographs Volume 3. IOS Press, Amsterdam, The Netherlands (2008)

38. Richter, K.F.: Prospects and challenges of landmarks in navigation services. In: Raubal, M., Mark, D.M. (eds.) Cognitive and Linguistic Aspects of Geographic Space–New Perspectives on Geographic Information Research, Lecture Notes in Geoinformation and Cartography, pp. 83–97. Springer, Berlin (2013)

39. Richter, K.F., Duckham, M.: Simplest instructions: Finding easy-to-describe routes for navigation. In: Cova, T.J., Miller, H.J., Beard, K., Frank, A.U., Goodchild, M.F. (eds.) Geographic Information Science. Lecture Notes in Computer Science, vol. 5266, pp. 274–289. Springer, Berlin (2008)

40. Richter, K.F., Klippel, A.: Before or after: Prepositions in spatially constrained systems. In: Barkowsky, T., Knauff, M., Ligozat, G., Montello, D.R. (eds.) Spatial Cognition V. Lecture Notes in Artificial Intelligence, vol. 4387, pp. 453–469. Springer, Berlin (2007)

41. Richter, K.F., Winter, S.: Citizens as database: Conscious ubiquity in data collection. In: Pfoser, D., Tao, Y., Mouratidis, K., Nascimento, M., Mokbel, M., Shekhar, S., Huang, Y. (eds.) Advances in Spatial and Temporal Databases. Lecture Notes in Computer Science, vol. 6849, pp. 445–448. Springer, Berlin (2011)

42. Richter, K.F., Winter, S.: Harvesting user-generated content for semantic spatial information: The case of landmarks in OpenStreetMap. In: Hock, B. (ed.) Proceedings of the Surveying and Spatial Sciences Biennial Conference 2011, pp. 75–86. Surveying and Spatial Sciences Institute, Wellington, NZ (2011)

43. Sadeghian, P., Kantardzic, M.: The new generation of automatic landmark detection systems: Challenges and guidelines. Spatial Cognit. Computat. 8(3), 252–287 (2008)

44. Schlieder, C.: Reasoning about ordering. In: Frank, A.U., Kuhn, W. (eds.) Spatial Information Theory. Lecture Notes in Computer Science, vol. 988, pp. 341–349. Springer, Berlin (1995)

45. Sorrows, M.E., Hirtle, S.C.: The nature of landmarks for real and electronic spaces. In: Freksa, C., Mark, D.M. (eds.) Spatial Information Theory. Lecture Notes in Computer Science, vol. 1661, pp. 37–50. Springer, Berlin (1999)

46. Surowiecki, J.: The Wisdom of Crowds. Doubleday, New York (2004)

47. Tezuka, T., Tanaka, K.: Landmark extraction: A web mining approach. In: Cohn, A.G., Mark, D.M. (eds.) Spatial Information Theory. Lecture Notes in Computer Science, vol. 3693, pp. 379–396 (2005)

48. Tomko, M., Winter, S.: Describing the functional spatial structure of urban environments. Comput. Environ. Urban Syst. 41, 177–187 (2013)

49. Waller, D., Loomis, J.M., Golledge, R.G., Beall, A.C.: Place learning in humans: The role of distance and direction information. Spatial Cognit. Computat. 2(4), 333–354 (2000)

50. Williams, S.: Generating pitch accents in a concept-to-speech system using a knowledge base. In: Proceedings of the 5th International Conference on Spoken Language Processing. Sydney, Australia (1998)

51. Winter, S.: Modeling costs of turns in route planning. GeoInformatica 6(4), 345–361 (2002)

52. Winter, S.: Route adaptive selection of salient features. In: Kuhn, W., Worboys, M., Timpf, S. (eds.) Spatial Information Theory. Lecture Notes in Computer Science, vol. 2685, pp. 349–361. Springer, Berlin (2003)

53. Winter, S., Raubal, M., Nothegger, C.: Focalizing measures of salience for wayfinding. In: Meng, L., Zipf, A., Reichenbacher, T. (eds.) Map-based Mobile Services: Theories, Methods and Implementations, pp. 127–142. Springer, Berlin (2005)

54. Winter, S., Richter, K.F., Baldwin, T., Cavedon, L., Stirling, L., Duckham, M., Kealy, A., Rajabifard, A.: Location-based mobile games for spatial knowledge acquisition. In: Janowicz, K., Raubal, M., Krüger, A., Keßler, C. (eds.) Cognitive Engineering for Mobile GIS (2011). Workshop at COSIT'11

55. Winter, S., Tomko, M., Elias, B., Sester, M.: Landmark hierarchies in context. Environ. Plann. B: Plann. Des. **35**(3), 381–398 (2008)

56. Zhang, X., Mitra, P., Klippel, A., MacEachren, A.: Automatic extraction of destinations, origins and route parts from human generated route directions. In: Fabrikant, S., Reichenbacher, T., van Kreveld, M., Schlieder, C. (eds.) Geographic Information Science. Lecture Notes in Computer Science, vol. 6292, pp. 279–294. Springer, Berlin (2010)

57. Zhang, X., Mitra, P., Klippel, A., MacEachren, A.M.: Identifying destinations automatically from human generated route directions. In: Proceedings of the 19th ACM SIGSPATIAL International Conference on Advances in Geographic Information Systems, GIS, pp. 373–376. ACM, New York (2011)

58. Zheng, Y.T., Zhao, M., Song, Y., Adam, H., Buddemeier, U., Bissacco, A., Brucher, F., Chua, T.S., Neven, H.: Tour the world: Building a web-scale landmark recognition engine. In: IEEE Conference on Computer Vision and Pattern Recognition, CVPR, pp. 1085–1092 (2009)

Chapter 6
Communication Aspects: How Landmarks Enrich the Communication Between Human and Machine

Abstract Landmarks are fundamental in human communication about their environments. This chapter will discuss what it takes to incorporate them into human-computer interaction. We will look at the principle requirements for such communication, and discuss how computers may produce and understand verbal and graphical references to landmarks. We will also present some results of studies testing the advantages of landmarks in human-computer interaction. We will see that there are huge benefits to gain from this integration, but also that there are still issues that need to be resolved.

6.1 Landmarks in Human-Computer Interaction

Throughout this book, we have established that landmarks are a key construct for humans to make sense of the environment they live in. Landmarks structure our mental representation of space and they are an important element of any spatial communication, be it verbal or graphical. Accordingly, producing and understanding references to landmarks comes natural to us. For a computer, this is far less straightforward; in fact, it is a rather hard problem as we have already argued in the introduction. Nonetheless, because it is so natural for us, enabling computers to use landmark references in their communication as well will lead to more natural, easier, and more successful human-computer-interaction.

In Sect. 1.3, we have exemplified what such a communication takes. This chapter will look at some approaches that aim at enabling it. Following Turing's ideas of a machine communicating with people without being identified as a machine [62], or more specifically, doing so in a spatial context [66], the aim here is not to copy the human cognitive processes and facilities. Rather, it is sufficient that the surface—the *interface*—of the computer's internal processes matches with human expectation and concepts. Further, it does not make sense to recreate human imperfection, biases,

K.-F. Richter and S. Winter, *Landmarks: GIScience for Intelligent Services*,
DOI 10.1007/978-3-319-05732-3_6, © Springer International Publishing Switzerland 2014

and mistakes in the human-computer interaction [31]. Rather, the machine should always pick what is (objectively) correct and best—a perfect super-human so to speak.

In the remainder of this chapter, we will first address ways of how machines may produce references to landmarks (in Sect. 6.2) because this appears to be easier than understanding references to landmarks, which we will discuss in Sect. 6.3. Finally, in Sect. 6.4, we will point to some studies that have investigated the usability and usefulness of introducing landmarks into human-computer interaction.

For a machine, it is easier to successfully produce landmarks than to understand them for computational and cognitive reasons. Computationally, a machine may rely on the data structures and algorithms presented in Chaps. 4 and 5—given that sufficient data is available. As we have discussed, this is hardly ever uniformly the case for any given environment (see also the final discussion in the next chapter). Accordingly, some fallback strategies in communication are required in case no landmark is available.

These computational methods ensure that the machine will pick landmark references that are salient and relevant. The pick may not necessarily be actually optimal though, since data may not be complete and any computational approach out of necessity makes some simplifying assumptions. However, generally this will not be an issue, since in contrast to machines people are very good at adapting to their communication partner. Simply by mentioning a geographic object in the communication, the communication partner will pay attention to this object, making it more relevant and, thus, increasing its salience [57]. The object becomes a landmark by virtue of being mentioned in the communication. This will compensate for the machine potentially picking not the most suited landmark reference. People will still be able to recognize and understand the chosen one in most cases. So, in short, in producing landmark references, machines can rely on a well-defined, but likely incomplete, set of landmark candidates to choose from—determined by the underlying data sets and algorithms—and on the cognitive facilities of the human communication partner, which enable them to flexibly adapt to the chosen landmark references.

On the other hand, this human flexibility and ability to adapt is the reason why understanding landmark references is so difficult for a computer. Even though we can assume a limited context in which interaction happens, namely negotiating some spatial descriptions (e.g., where something is located or how to get to some place), the variety of geographic objects people may select from and the variety in the ways they may describe this selection is immensely wide. This means that the computer needs much greater flexibility compared to the case of producing landmarks, and machines are not very good at adaptation. Thus, either the vocabulary needs to be restricted for successful understanding of landmark references—which takes away many advantages of landmark-based communication—or a sophisticated mechanism for resolving (arbitrary) landmark references is required, including detailed data about the environment and advanced parsing functionality. Alternatively, machine learning mechanisms may be employed to learn from communication with a human partner; however, to date we are not aware of any approach actually following down that path.

In short, it is much harder for the computer to infer what kind of landmark reference a human interaction partner will produce than for the human to understand machine generated landmark references. This makes understanding landmark references the more difficult challenge.

6.1.1 Requirements for Machine Spatial Communication

First, let us have a look at a sample dialog requesting directions to a specific location between two people. We have already discussed the principles underlying such dialogs in Sect. 3.4 of this book.

Suppose you are in Melbourne, Australia, for the first time. You have been wandering around the Central Business District (CBD) for a while, have ended up in front of the exhibition centre just south of the Yarra river, and now you want to go back to your hotel, which happens to be at the eastern end of Little Bourke Street (one of the streets of the CBD, and the eastern end being part of Chinatown). Luckily, there just happens to be one of the friendly guides provided by the City of Melbourne for tourist support, hanging around here. So you decide to approach them and ask for directions. The dialog may go something like this:

You: Excuse me?
Guide: Yes?
You: How do I get to Chinatown?
Guide: Chinatown? You mean Little Bourke Street?
You: Yes.
Guide: Ok. You cross the street here (points east) *and walk along the river past the casino. Then you cross another big street—Queens Bridge. Just behind that street there is a footbridge crossing the river. Take this and continue along the river on the other bank until you get to Flinders Street Station. Do you know that?*
You: Nod, because you have passed it earlier today and remember it from the travel book.
Guide: Good, you can go either through the tunnel underneath the station or get onto the bridge just next to the station and then turn left away from the river. This will get you onto Swanston Street, which you will probably know as well. Just walk up that street for a bit and Little Bourke Street will be on your right. You can easily spot it by its Chinese gate at the entrance to the street.
You: Oh, that sounds easy. I just walk along the river here, cross it and then go up Swanston Street. Thank you very much!
Guide: Yes, exactly. Have a nice day!

You may have experienced many similar dialogs in your life. As discussed in Sect. 3.4, we can observe several phases in this example dialog, namely a initiation phase ('Excuse me?'), the route directions themselves ('You cross the street here...'), a securing phase ('Oh, that sounds easy'), and a closure phase

('Have a nice day') [1, 15, 18, 49]. In each phase, each communication partner has different tasks, and both need to produce and understand references to landmarks.

In human-computer interaction there is, of course, no need for social conventions—the computer will always behave the same way. In the following we will assume that the human user requests some spatial information (location of a geographic object, or route directions) from the computer. We also assume that the context is defined, i.e., we do not assume a general purpose query machine, but some service dedicated to spatial communication, for example, a navigation service.

The requirements for a computer to produce landmark references have been discussed throughout Chaps. 4 and 5. They can be summarized as follows:

1. A data set (or the combination of several data sets) about a geographic environment that contains data which makes geographic objects discriminable from each other and allows to access (some of) their properties (such as location, size, or type).
2. A data structure that allows for capturing the landmarkness of geographic objects and for integrating this information into other required data (e.g., a path network for routing purposes).
3. A mechanism to establish an object's principle suitability as a landmark—its *salience*.
4. A mechanism to establish an object's suitability in a specific given situation—its *relevance*.

To make landmark production work it needs the actual communication mechanism(s), i.e., some way to communicate the chosen landmark reference(s) to the human user in verbal or graphical form in the right context.

In short, the production of landmark references by a computational system seems to be a reasonable expectation. However, people may have differing previous knowledge about an environment and, thus, require different detail in the spatial descriptions. Or they may have some clarification questions (in a *securing* phase) and the system would need to understand and come up with alternative or more detailed descriptions.

Understanding landmark references is the greater challenge compared to producing them. Accordingly, the requirements for successfully understanding landmarks may be greater as well. At least they appear to be less well defined. The minimal requirements are:

5. A mechanism to parse human queries (verbal or graphical) into a form that the computer can process.
6. A data repository that allows for a matching of landmark references to geographic objects—something like georeferencing [36], only that the references can be expected to be less well specified than in an authoritative gazetteer.
7. Some (spatial) reasoning mechanisms to correctly relate all references in the query, which then allows to come up with a reasonable answer—bringing us back to producing landmark references.

6.2 Producing Landmarks

In their initial paper on landmark identification, Raubal and Winter [54] have already suggested a formal notation for including landmarks into route directions. Approaches for producing landmarks are to this day predominantly developed for enriching route directions with landmarks. Most notably, the Australian routing service WhereIs augments its route directions with landmark references that are derived from an external collection of POIs (see Chap. 5).

As a little exercise you may try to recreate yourself the route described in above dialog. The best route we could achieve on WhereIs contains some unnecessary turns and loops and, thus, results in 22 instruction steps. This is clearly much more complicated than the human dialog. For that reason, we will not use this particular route for the following example.

6.2.1 Producing Verbal Landmark References

Commercial routing services usually generate their instructions using some pre-defined templates with little linguistic variation. The same holds for WhereIs, which uses a template of the form "<TURN ACTION onto STREET NAME at POI NAME>" to incorporate landmark references into the generated instructions. For example, when calculating directions from the Melbourne Exhibition Centre (on Normanby Rd, Southbank) to The University of Melbourne (156–292 Grattan St, Parkville),[1] five (out of the 11) instructions contain a reference to a POI landmark: "At the roundabout, take the 1st exit onto Peel St, West Melbourne at *Queen Victoria Market*"; "Veer left onto Elisabeth St, Melbourne at *Public Bar Hotel*"; "At the roundabout take the 3rd exit onto Elisabeth St, Melbourne at *Dental Hospital*"; "Turn right onto Grattan St, Melbourne at *Royal Melbourne Hospital*"; "Turn left onto UNNAMED road after Barry St, Carlton at *The University of Melbourne*". As you can see, these instructions utilize large, prominent buildings as landmarks, which are easy for wayfinders to identify. At the same time the generated instructions are very monotonous, especially given that those instructions without landmark references (not printed here) look very much the same without the "at" part.

In research, there have been several attempts of breaking up these monotonous directions with more variation, using methods from natural language generation (NLG). Dale et al. [17], for example, argued for a more natural sentence structure

---

[1]Like most routing services, WhereIs requires address information to actually calculate a route.

Fig. 6.1 The architecture of
the CORAL system
(after [17])

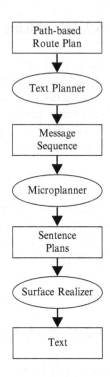

that gets away from the one-sentence-per-step messages routing services produce, and instead make use of more complex clause structures that cover related information in a single sentence. Their CORAL system generates such instructions; its architecture is shown in Fig. 6.1. This kind of architecture is quite typical for NLG-motivated route planners and reflects to a degree the theoretical models for giving route directions discussed in Sect. 3.4.

CORAL starts out with a route through a network representation of the environment, very similar to what is used in any routing service. It then moves through three phases to arrive at the verbal messages communicated to the user. The first phase, the *text planner*, determines what information about the route needs to be conveyed. This information is represented using three different message types: POINT messages refer to landmarks (e.g., 'turn left *at the gas station*'); DIRECTION messages correspond to turns made at decision points (e.g., 'turn right'); PATH messages describe continuous movement along parts of the road network (e.g., 'follow the road'). The text planner phase results in an alternating sequence of POINT, DIRECTION, and PATH messages, finishing with a POINT message that describes the destination.

This text plan is the input for the *micro-planning* phase, which decides how to combine the individual messages into clause structures and also how to refer to each of the referenced elements. The former is an aggregation task, where two

Table 6.1 Example output of
the CORAL system;
from [17]

Start at Parbury Lane
Follow Parbury Lane until you reach the end
Take a right
Follow Lower Fort Street for 30 m
Turn to the left at George Street
Follow George Street until you reach your destination

(or more) messages are merged. Most typically, this is a PATH + POINT (e.g., 'follow the road until the roundabout') or POINT + DIRECTION (e.g., 'turn left at the gas station') combination, but other combinations are possible as well. The latter part of the micro-planning phase is generating referring expressions—including selection of the reference noun (or pronoun, or other construct) used to describe an intended object, such as an intersection or landmark. Following the principles of relevance [33], the CORAL system tries to be as specific as needed, following the three principles of sensitivity, adequacy, and efficiency. It in turn uses as a reference alternatively either a landmark that is at or close to an intersection, the type of intersection (e.g., T-intersection or roundabout), the name of the immediately preceding intersection, or the name of the intersecting street.

Finally, the *surface realizer* phase maps the semantic specifications into actual sentences, i.e., into (grammatically correct) natural language. An example output of the CORAL system is shown in Table 6.1; this example is taken from [17].

As you can see, these instructions have some more linguistic variation than the WhereIs directions discussed above (on p. 179). For example, turns are described by stating 'take a right' or 'turn to the left'. Curiously enough however, the particular example provided by Dale et al. does not contain any landmark references, with an exception of the structural landmark 'the end of the road' [42].

Another system that aims for variation in the generated route directions is an information kiosk developed at the University of Bremen, Germany [16]. While set in an indoor scenario, it makes explicit use of landmark references and is based on some of the methods discussed in Chap. 5. There is no principle reason why it would not work in an outdoor setting as well. Figure 6.2 shows an overview of the pipeline architecture employed in the kiosk system. For now we are only interested in the bottom part, i.e., the pathway starting at the 'spatial data' box and leading to the 'user' box. We will discuss the top part in the next section.

Except for being a dialog system, i.e., enabling some synchronous, co-presence user interaction, the system's approach to producing route directions is similar to the CORAL system. The kiosk system starts by calculating a route between the kiosk itself, which always is the origin since users are standing in front of the computer running the kiosk software, and the requested destination. Next, the instructions required to successfully navigate this route are determined. To this end, a computational process called GUARD (Generation of Unambiguous, Adapted Route Directions) [57] is employed. GUARD unambiguously describes a specific

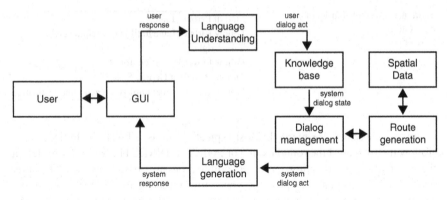

Fig. 6.2 Pipeline architecture of the kiosk system (after [16])

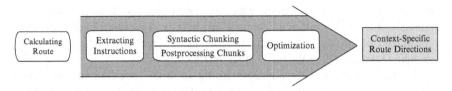

Fig. 6.3 Overview of GUARD, the generation process for route directions in the kiosk system (from [57], modified)

route to a destination with the instructions adapted to environmental characteristics. The generated route directions make heavy use of landmark references to structure the instructions. Figure 6.3 provides an overview of the GUARD process.

First, for every decision point every possible instruction is generated (e.g., 'turn left', 'turn left after the glass door', 'turn left at the posters' may all be possible descriptions for a left turn at a corridor intersection where you have just passed a glass door and there are posters attached to the wall). This results in a set of possible instructions for each decision point. Clearly, given the principles of relevance, the system cannot use all of these possible instructions, but has to decide on one of them. As an intermediate step for getting to this decision, GUARD performs *spatial chunking* [41, 43] in a second step. This is similar to the *aggregation* step in the CORAL system in that spatial chunking combines several consecutive instructions into a single one (e.g., 'turn left at the third intersection' instead of 'go straight', 'go straight', 'turn left'). The resulting spatial chunks are possible clause structures in the CORAL terminology.

But again, GUARD produces every possible chunk that is deemed valid. Thus, next the system needs to decide on the 'best' description for the route, i.e., the optimal sequence of chunks. This is an optimization process. The resulting chunk sequence depends on the chosen optimization criterion. Typically, GUARD aims for compact route directions, i.e., chooses the minimal number of chunks to cover the whole route. References to landmarks are integrated into these instructions already in GUARD's first step, employing the landmark handling approach by Richter [56] discussed in Chap. 5. In fact, landmarks are an important element of spatial chunking (see [43] for an in-depth discussion). They enable structuring route directions into chunks, such as 'follow Parbury Lane until you reach the end' (from the CORAL example above) or 'turn left at the church' (which may still be a long way down the road).

The chunk sequence is used as input for the language generation module (via the dialog management, but we can ignore that for now). The language generation module is a full-fledged NLG system based on the pCRU framework [5]. This framework—probabilistic context-free representational underspecification is its full name—allows for resolving the issue that getting from a semantic representation, such as the chunks produced by GUARD, to a specific linguistic expression is almost always underspecified. There are many ways to express a change of direction to the left, for example, 'turn left', 'turn to the left', 'make a left turn', 'left', or 'go left'. With pCRU, these variations can be formalized in a context-free grammar,[2] and each possible variation can be assigned a probability of being generated by the system. The kiosk system specifically uses landmarks as (intermediate) destinations along the route. The route is segmented using landmarks along the way. For example, the system may produce the following instructions to reach a particular office: 'Turn around and go straight until the first corridor on your left. Turn left and go straight until door B3180 is at your left.' As you can see, these instructions are similar in structure to the CORAL instructions listed in Table 6.1, but have at least one landmark reference in every instruction step.

Dethlefs et al. [20] also use pCRU in their route direction generation approach. They create different kinds of direction—turn-by-turn directions for unfamiliar environments and destination descriptions for familiar ones. In that, they combine and adapt the approach to determining the salience of streets by Tomko et al. [61] and the idea of landmark categories by Duckham et al. [22] to determine the content of the route directions, and then use pCRU and aggregation mechanisms to create natural language route directions with linguistic variation in the instructions. Table 6.2 lists examples for turn-by-turn instructions and destination descriptions generated by their system for a route from the Richmond South Post Office to the Richmond Cricket Ground (Richmond is a suburb of Melbourne, Australia).

---

[2] We will not explain context-free grammar any further in this book other than saying that it is a concept of formal language theory. A grammar essentially is a mapping from some nonterminal symbols (e.g., the semantic representation of a left turn) to some (string of) terminal symbols (e.g., the words 'left', or 'go left'); the grammar is context-free if a specific nonterminal symbol V always maps to the same string of terminal symbols w, regardless of its surrounding symbols (the context). For more details on formal languages refer to a textbook on theoretical computer science or theoretical linguistics, for example, Hopcroft, J.E., Motwani, R., Ullman, J.D. (2006). Introduction to Automata Theory, Languages, and Computation (3rd ed.). Addison-Wesley.

Table 6.2 Route directions generated by the system of Dethlefs et al., taken from [20]

Turn-by-turn directions
Follow Swan Street to Richmond South Post Office for 52 m. Take a left after the Richmond South Post Office, and take a right at Woodlawn Street. At this point, go past a car park on the left. Cross the next junction at Normanby Place, and cross the next junction at Moorhouse Road. This will then lead onto Wellington Parade. Go straight over junction. Cross the next junction, when you get to Vale Street. Go into East Melbourne. Take a slight left after the East Melbourne Post Office. At this point, go past traffic lights on the right. Head straight over junction, and take a turn after the traffic lights. Take a left before the Barbecue, and after the Barbecue, take a slight right. At this point, pass the Richmond Cricket Ground on the left and continue for 62 m

Destination descriptions
Head to the East Melbourne Post Office that is on Wellington Parade. Take Rotherwood Street and then go over Wellington Parade. You will come to the Richmond Cricket Ground

You can observe that there is a lot of linguistic variation in the turn-by-turn directions. They are also much more verbose than those of the CORAL or kiosk systems. Indeed, people had some difficulties realizing that these route directions were machine-generated (see Sect. 6.4 for further discussion). You can also see that the system may refer to slightly odd landmarks, such as a barbecue. Public barbecues are placed in many parks in Australia, but they are significantly smaller than, say, buildings, making them rather unlikely candidates for landmark references in route directions for longer routes.

This last example, and also the example of the CORAL system at the beginning, highlight again that computational systems depend on data sets that contain a sufficient number of potential landmark candidates in order to produce useful references to landmarks sufficiently often.

A simple way of dealing with this data issue is to restrict the system to a specific location. The SpaceBook project, for example, develops a tourist information system with pedestrian navigation functionality for the city of Edinburgh [38]. The system integrates navigation instructions and the provision of tourist information about relevant POIs into a single dialog. Navigation instructions incorporate references to landmarks—both the tourist POIs as well as other salient buildings, such as restaurants. The data is taken from a city model spatial database that contains information about thousands of objects in Edinburgh (according to the authors). The data has been compiled from existing sources, such as OpenStreetMap, Google Place[3] and the Gazetteer for Scotland.[4] More details about SpaceBook will be presented in the next sections of this chapter.

---

[3]https://www.google.com/business/placesforbusiness/, last visited 8/1/2014.
[4]http://www.scottish-places.info/, last visited 8/1/2014.

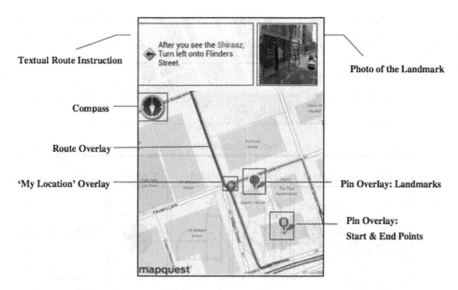

Fig. 6.4   A prototype multimodal route guidance interface [70]

6.2.2   Producing Graphical Landmark References

Now let us turn to producing landmarks in a graphical form. Unfortunately, there has not been much work done on a dedicated production of graphical landmark displays. Commercial (in-car) navigation services often depict POIs on their maps, or at least offer the option to do so. However, as argued before in this book, they cannot really function as landmarks. Curiously, while WhereIs produces references to landmarks in its verbal instructions, the accompanying route map does not highlight these landmarks. The Bremen kiosk system in one of its iterations depicted landmarks as part of a multi-modal presentation of route information. But this has been a rather ad-hoc solution basically labeling those polygons that represent objects mentioned in the description, while not labeling other object polygons. This research has never been published or pursued any further. There has been some preliminary unpublished work in relation to the OpenLandmarks idea [70], where both the location of a referenced landmark on a moving-dot map as well as a photograph of that landmark were shown as part of multi-modal instructions (see Fig. 6.4). Again, this is work in its earliest stages.

Elias et al. [26] discussed adequate depictions of landmarks on maps. This is a general discussion not regarding an attached landmark production system. Their analysis is from a cartographic perspective and provides some guidelines for including landmarks on a map. In their analysis, they distinguish different ways of referring to a building, namely either using a shop's name (McDonalds, H&M,

Table 6.3 Design guidelines for producing graphical landmarks (after [26])

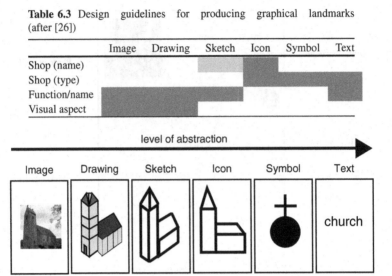

	Image	Drawing	Sketch	Icon	Symbol	Text
Shop (name)						
Shop (type)						
Function/name						
Visual aspect						

level of abstraction

Image   Drawing   Sketch   Icon   Symbol   Text

church

Fig. 6.5 Levels of abstraction in depicting landmarks on a map (after [26]). The photograph in the first panel is CC-BY-2.0 by Flickr user *Michiel2005*, modified

Starbucks, etc.) or its type (gas station, pharmacy, bakery, etc.), or the function of a building (school, church, etc.) or their visual properties (the big red building, the small wooden shed, etc.). Table 6.3 shows the results of their analysis using several levels of abstraction (from image to textual description) illustrated in Fig. 6.5.

The matrix in Table 6.3 highlights the most suitable way of graphically presenting a landmark. According to Elias et al. [26], a textual description would always be possible, but is not always adequate. For shops of well known brands, the brand's icon may be the ideal representation as it will be easily identifiable for most people. For shop types specifically designed icons or symbols may be used (e.g., a stylized bank or gas station). Buildings with specific function are often large and may be at (structurally) prominent locations, thus, showing them in some detail—or at least their outline—helps in identifying them in the real world. Visual aspects are best represented directly, i.e., by using a (photographic) image or a drawing that highlights this visual property. To convey proper names of buildings, a textual label has to be used.

There has been some research on using photographs for augmenting wayfinding assistance, which implies the use of landmark-based navigation concepts even though this is not explicitly discussed. For example, Hirtle and Sorrows [37] presented a library finder for the University of Pittsburgh campus. This system uses a hierarchical approach. It first presents an overview of the campus with the library indicated and then offers more detailed information on how to get there (to the building and then within the building). Each of these hierarchy levels is accompanied by a single photograph that highlights the most important visual information for finding the way, for example, the entrance to the building. Lee et al. [48] integrated photographs of landmark buildings in a map view of a route. In both systems, the authors themselves collected the necessary photographs. There is no automatic identification of potential landmarks. Also, given the time of their development (late 1990s and early 2000s, respectively), both approaches run as desktop web browser application, and are clearly outdated in their technology.

Only a few years later, the first mobile applications using photograph-based navigation appeared (e.g., [32, 44]). Beeharee and Steed [3, 4] used photos to augment wayfinding instructions presented on mobile devices (here, personal digital assistants, or PDAs for short). While this clearly made the system more useful, since now users can directly compare what they see in the environment with what they see on the photograph, photographs are still selected by the authors. There is no indication how this system might scale up, even though, the authors state that technically it would work anywhere (in the UK). You can clearly observe the transition to today's modern smartphone applications, but a lot remains rather awkward in this early system, though in some sense it was ahead of its time.

Hile et al. [34] presented another photograph-based navigation service for mobile phones. The system uses a database of previously geotagged photographs from an outdoor augmented reality project [60], which contains photographs of mostly the Stanford University campus. The system segments an environment into loxels, which are 30×30 m grid cells. For each loxel the current route runs through, the system determines all photographs visible from that loxel. Based on nearness to the route and deviation from movement direction a cluster of photographs gets selected with each cluster containing a set of photographs with similar views. The center of the cluster provides a canonical view, similar to those approaches discussed in Sect. 5.2.2. This canonical view is presented to the users along with textual instructions on how to proceed along the route. Additionally, an arrow is overlaid on the photograph indicating the direction to take.

Similar to the system of Beeharee and Steed, the system of Hile et al. faces the issue of how to provide a sufficient number of useful photographs. This links back to our discussion about crowdsourcing as an alternative approach to data collection (see Sect. 5.5.3).

Recently, Microsoft Research presented a pedestrian guidance system using street level panorama images [68]. The system is based on Nokia City Scene,[5] which combines panoramic photographs, city models constructed from LIDAR data, and map data of about 20 of the major cities around the world. City Scene is similar to Google Street View, but with significantly less coverage. It seems to focus mainly on the USA, and at the time of writing it is restricted to the Nokia N9 smartphone.

The system highlights specific features as landmarks, primarily business signs. In case no signs are available, a building is picked instead. A text recognition pipeline is able to automatically extract signs from the photographs determining for the

---

[5]http://store.ovi.com/content/178183, last visited 8/1/2014.

approaching direction which sign is the most visible at an intersection. It is not sufficient to do this simply from the middle of the street, but it needs to account for which side of the street a pedestrian is traveling on. Accordingly, the algorithm accounts for the orientation of a sign relative to the orientation of the street to ensure it is actually visible. The fallback solution of selecting a building only accounts for a building's location (the corner of and the distance to the intersection). Conceptually, Wither et al.'s approach is similar to the approach of Raubal and Winter presented in Chap. 5.

The system presents panoramic images at decision points, with an enlarged display of the sign at the bottom of the screen to ensure it is readable on screen. These panoramic images are canonical views of an approaching intersection selected for eight different directions, i.e., for each building eight canonical views are calculated. Users can click back and forth between panoramas, but cannot otherwise alter the view. Based on a user's location, the system automatically selects the next relevant panorama.

In addition to the common problem of huge data demands—in this case Navteq collecting the required city models and photographs, the Nokia system is mainly useful for business districts and inner cities as it heavily relies on (business) signage on buildings. This renders the system, which is technically promising, insufficient for more global use as a navigation service.

6.3 Understanding Landmarks

After we have discussed how computers can produce landmark references, we now turn to how they may be enabled to understand references to landmarks. As discussed in the introduction to this chapter, this is a much harder problem than producing such references.

6.3.1 Understanding Verbal Landmarks

Understanding verbal references to landmarks is first and foremost a matter of natural language processing (NLP). There has been a lot of progress in NLP over the years. For an overview of the state of the art see, for example, the book by Jurafsky et al. [39]. We can assume that the parsing of written and spoken natural language input will soon be reliable. This does not provide a full interpretation of what has been said yet, however. Individual words can be identified and also how they relate to each other in the uttered sentence. For example, the utterance 'I am at the bar in the cinema' places the speaker in something called 'bar' which is located in something called 'cinema'. To get the full meaning, the computer needs to understand what 'bar' and 'cinema' means—here ontologies come into play [2].

Then it is important to understand which specific bar and cinema are referred to.[6] This requires also an understanding of the context of the utterance [53].

WhereIs solves this problem by forcing users to resolve ambiguities themselves. It is possible to enter POI names as origin and destination (in the 'what' text box), however, the user needs to additionally provide the suburb of the POI (in the 'where' text box). Then WhereIs suggests a list of potential street addresses that the input may refer to and makes the user select the appropriate one. This ensures that WhereIs gets the context right, but on the other hand forces users to provide information they possibly do not know.

Such context restrictions are typical for computational services. The above argument is not a specific criticism of WhereIs, other well-known (web-based) navigation services behave similarly. For example, Winter and Truelove [67] have shown that while Google Maps allows for more free-form input compared to WhereIs, interpretation of this input is often inadequate and requires users to either adapt their input to the system interpretation or to put significant effort in interpreting the results themselves. For example, Google Maps currently still interprets any spatial relation as 'near'.

Although Winter and Truelove's analysis was drawn from the interpretation of place descriptions, which are structurally different from landmark references as shown before, many findings also hold for understanding landmark references. Also in the context of place description research, Vasardani et al. [64] highlighted the following issues:

- Official, authoritative gazetteers do not usually include unofficial and vernacular place names, or temporary (replacements of) place names, such as event names that are sometimes used synonymously for the location they are happening at. This is in contrast to the popular use of vernacular place names, in particular in familiar environments. Sometimes people may not even be aware of official names. Thus, the restriction to official place names restricts a system's understanding of references to places and in turn restricts interaction possibilities with the users.
- A sensible interpretation of spatial relations is important to understand place references correctly. Clearly, taking every relation to mean 'near' is inappropriate. Formal models for a range of spatial relations have already been investigated in the literature (see Chap. 4). In principle, these models enable a more adequate interpretation of place references, however, the interpretation of spatial relations is context-dependent [69]. Also, people's cognitive concepts of a relation may differ from what is defined in a formal model [40, 52] leading to diverging interpretations of a place description.
- Places often are extended geographic regions with indeterminate boundaries. This indeterminacy is hard to capture satisfactorily in computational models, despite of several attempts [6, 12, 47]. Again, context and individual differences aggravate the issue.

---

[6]Even though in this example it may be sufficient to know which cinema is meant, assuming there is only one bar in it.

The latter two issues are important aspects of understanding locative expressions, as has been discussed in Chap. 4. The first relates back to the context and content restrictions. Such restrictions are unavoidable as no computational system will ever understand every possible landmark reference possible (globally, i.e., around the world, or even locally, e.g., within a city). This is a big difference to a human communication partner who will have no problems dealing with vernacular or unofficial references, or with references to vaguely defined geographic objects, though there might be occasional misunderstandings. Not being able at all to deal with such indeterminacy is a major restriction for computational systems.

Coming back to the systems already presented in Sect. 6.2.1, for some systems solving these issues is still relatively easy. The kiosk system [16], for example, is stationary and placed inside a building. Accordingly, meaningful dialogs can only be expected about this specific building. And for a single building, even if it is large, it is quite reasonable to manually collect all relevant data needed for landmark-based communication. Most importantly, such data will comprise of rooms (their number, function and occupation), some infrastructure, such as coffee machines or printers, and some landmark objects, for example, poster walls or furniture placed in corridors (cf. also [58]). This data will then form the system's knowledge base. The kiosk system uses it to both produce and understand landmark references (see Fig. 6.2 on p. 182).

The kiosk's dialog manager uses different dialog states (Table 6.4) and a deterministic dialog policy defined in Eq. (6.1).[7] The sequences in Eq. (6.1) (e.g., 10,000,000) are to be read against Table 6.4. Digits in the sequence represent the domain value of each state, in the order the states appear in the table. That is, $10, 000, 000$ corresponds to a state where a greeting has happened and the system awaits a user request.

Essentially, the dialog manager proceeds through a dialog according to the previously presented communication models. It reacts to utterances by users, which accordingly have to be parsed in order for the system to provide useful replies. The kiosk system relies on textual input, so typed utterances are being parsed. To this end, the OpenCCG parser is used [14], producing a structured representation of the utterances' semantics, which is matched against the knowledge base. In case the parser fails, keyword spotting is used to look for names of locations or people, which may help in guessing the user request.

---

[7]In fact the dialog manager uses a Markov Decision Process (MDP) model, but this is not really important here. For more details on MDP, look up a textbook on Artificial Intelligence, for example, S. Russel & P. Norvig, Artificial Intelligence: A Modern Approach, 3rd ed. Prentice Hall, Englewood Cliffs, NJ.

Table 6.4 Dialog states used in the kiosk dialog manager, after [16]

State	Domain value
Salutation	0 = null; 1 = greeting; 2 = closing
Origin	0 = unknown; 1 = requested; 2 = known
Destination	0 = unknown; 1 = requested; 2 = known
NumTuples	0 = null; 1 = one; 2 = more-than-one
Instructions	0 = unknown; 1 = known; 2 = provided
UserUtterance	0 = unknown; 1 = parsed; 2 = unparsed; 3 = spotted
MoreInstructions	0 = null; 1 = empty; 2 = yes; 3 = no

$$p(s) = \begin{cases} \text{opening} & \text{if } s \in \{0000000\} \\ \text{request} & \text{if } s \in \{1000000, 1000012\} \\ \text{other\_request} & \text{if } s \in \{1220210, 1220220\} \\ \text{query\_route} & \text{if } s \in \{1220210, 1220220, 1110030, 1210030, \\ & \quad 1220030, 1211030, 1221030\} \\ \text{present\_info} & \text{if } s \in \{1221110, 1221130\} \\ \text{clarify} & \text{if } s \in \{1112100, 1112030, 1212030, 1222211, 1222231\} \\ \text{apologize} & \text{if } s \in \{1110020, 1210020, 1220220, 1210210\} \\ \text{confirm} & \text{if } s \in \{1112010, 1112030\} \\ \text{closing} & \text{if } s \in \{1 * * * * * 3\} \\ \text{wait} & \text{otherwise} \end{cases}$$

$$(6.1)$$

In an user evaluation, this fallback to keyword spotting turned out to be extremely useful since the parser often failed to parse a sentence properly. There is a wide variety of ways of asking for a location, often with ungrammatical structure (something the parser does not take well), or words missing from the lexicon against which the parser operates. The keyword spotter, however, correctly identified the addressed location in almost all cases the parser failed, such that in the end only about 3 % of utterances were not understood by the kiosk system. This shows again that understanding landmark references is much harder than producing them, and that a comprehensive knowledge base is invaluable in understanding landmark references.

The SpaceBook tourist guide [38] also employs a geographic restriction of context supporting only requests about Edinburgh. Being mobile, the system also exploits users' current location and their pace. Based on this information, the system pushes potentially interesting POIs to the user ('near you is the statue of David Hume'). While users may still ask about other geographic objects or information about Edinburgh (and the system does not restrict this), pushing POIs makes it highly likely that users will request more information about them. This would offer another option of working around the issues listed above—in this case, however, this was not the authors' intention. POIs are being pushed to inform the user, a tourist, about interesting facts about Edinburgh.

The SpaceBook system uses commercial software for parsing natural language input as well as for speech generation. An interaction manager keeps track of the geographic context and the dialog history to resolve anaphoric references,[8] among others. Questions related to the navigation process are answered by computations within the system (e.g., distances between locations). Touristic questions are answered by a textual lookup from the Gazetteer of Scotland, Wikipedia[9] and WordNet.[10] A ranking mechanism, which was trained using machine learning techniques, ranks candidate answers and the top candidate is presented to the user. Since answers often involve fairly long text, they are provided piecewise and users can interrupt their presentation anytime. Understanding what kind of question is being asked is again achieved through machine learning, i.e., by training another classifier using an annotated question corpus.

6.3.2   Understanding Graphical Landmarks

Not many interfaces exist allowing for spatial input in graphical form, let alone graphical input related to landmarks. This may change in the (near) future with the advent of smartphones with their advanced gesture input recognition. There has been a long run of research on so called *query by sketch* [23], however. Query by sketch permits users to sketch (draw) a spatial query to request some information from a service. Such sketches are first and foremost a sequence of pixels, which may be grouped to lines (a geometric operation). These lines may then be further grouped to represent some geographic objects (a semantic operation, even though it will incorporate geometric sub-operations), such as streets, buildings, or labels. Figure 6.6 shows a fictitious sample sketch map that is quite typical for how people produce such maps in a communication act.

As discussed by Forbus et al. [28] it takes a big effort for a computational system to understand sketches. The authors characterize the sketching ability of a system as well as of its users along four dimensions: *visual understanding, conceptual understanding, language understanding, drawing skills.*

Visual understanding refers to the ability of making sense of the 'ink' used while sketching, i.e., how different line strokes form and how these line strokes in turn may form more complex objects (so called 'glyphs'). Geometric operations as mentioned above fall into this kind of understanding. Conceptual understanding helps translating these objects into meaningful elements (such as streets or

---

[8]Anaphoric references use some kind of deictic reference, usually a pronoun, to refer back to an item mentioned before. For example, in 'St Peter is a large cathedral in Rome; it is home to the Pope' 'it' is an anaphoric reference to St Peter.

[9]http://www.wikipedia.org, last visited 8/1/2014.

[10]http://wordnet.princeton.edu/, last visited 8/1/2014.

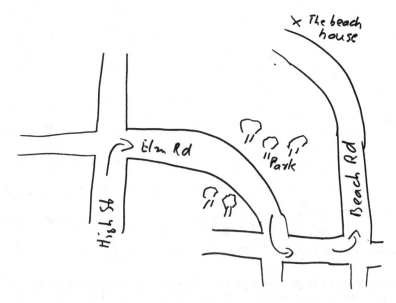

Fig. 6.6 A (fictitious) sketch map

buildings). Language understanding is particularly useful in multimodal settings, where the system also allows for verbal input. In human-human communication it is common that communication partners verbally explain their sketch while drawing. Multimodal input may also help the system in interpreting what is being drawn— given it has sufficient NLP capabilities. Finally, drawing skills determine how well a human or computer sketcher can translate their conceptual ideas into a recognizable drawing.

Visual understanding, i.e., the first processing of the sketched input, may start with simplifying the line strokes drawn by the user (e.g., by employing the well-known Douglas-Peucker algorithm [21]). Not all bends and curves will be intended by the user, some may just occur for reasons of motor skill or input resolution. The next step is to aggregate individual strokes to objects. According to Blaser and Egenhofer [7], for each new stroke it needs to be decided whether this stroke forms the beginning of a new object, whether it belongs to the current object, or whether it is an addition to an already existing object. The drawing sequence plays a crucial role here, i.e., the order in which strokes are produced and pauses made between strokes. Also, the location of the new stroke relative to other objects needs to be considered. Boundary zones around objects are used to decide whether a stroke is part of that object (if start or end point of the stroke fall within this zone) or not (see Fig. 6.7a).

Once an object has been identified it is further processed to clean up any missing links or to remove overshoots, and to determine whether the object represents a line, a region, or a symbol. For region objects (polygons), edges may be connected

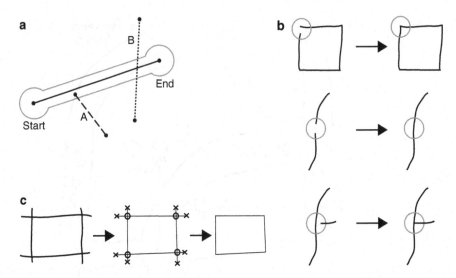

Fig. 6.7 Different geometric operations in processing a sketch, after [7]. (**a**) Buffer zones used to determine whether a stroke forms part of an object; (**b**) closing gaps within an object; (**c**) removing overshoots for polygonal objects

to close gaps, for lines interruptions in their flow may be closed (see Fig. 6.7b). If objects form a closed loop, i.e., represent a region, any overshooting lines are removed (see Fig. 6.7c).

Furthermore, visual understanding has to determine how the various objects relate to each other, i.e., to understand spatial relationships in the sketch. It is widely accepted that qualitative spatial relations are best suited to this end [13, 23, 29, 65]. Egenhofer [23] suggested to use five types of binary spatial relations: coarse topological relations (using the nine-intersection model [24]), detailed topological relations, metrical refinements (in line with [25]), coarse cardinal directions [30], and detailed cardinal directions. Using these relations results in fairly complex, comprehensive sketch descriptions that are also flexible enough to allow for relaxing relations in an ordered manner. This is important for querying spatial data based on sketches where users ask for configurations that reflect the provided sketch in a qualitative way, i.e., describe scenes that are similar to the one sketched.

After establishing what objects there are in a sketch and how they relate to each other, the next step is to understand what these objects represent. This *conceptual understanding* is much harder than *visual understanding*, which should not be hard to guess given the previous discussions around geometric and semantic information in this book.

The simplest way of handling conceptual understanding is to offer a fixed vocabulary. Such a fixed set of objects and relationships between these objects restrict what users may express in their sketch. In general, it helps understanding if the domain is restricted, thus, domain-specific systems have an advantage. But beyond domain-specific symbology, there are graphical conventions and standard visual symbols that help understanding a sketch. For example, stick figures typically represent humans, and sequences of snapshots often depict dynamics [28].

To illustrate let us look at the CogSketch system [29]. It is a sketch understanding system with an in principle open domain. For conceptual understanding, it relies on the user to label drawn objects. Users have to select a fitting concept from CogSketch's knowledge base, which is derived from OpenCyc[11] and contains over 58,000 concepts. Three different interface techniques have been explored: (1) concept lists providing access to a (small) subset of the knowledge base (selected for a particular application); (2) direct access letting a user type in a concept name and the system providing string completion; (3) glyph bars providing pictorial labels for knowledge base concepts. The latter is similar to the first interface mode, however, users do not need to know anything about the knowledge base, only to understand the glyphs.

With visual and conceptual understanding, it becomes possible for a service to interpret a user's sketched input and to interact by providing reasonable answers. The original query-by-sketch application by Egenhofer [23] translated the conceptual understanding of the sketch (represented as a semantic network in their case) into SQL (or SQL-like statements) for database lookups. Sketching here was seen as an alternative, more user-friendly interaction mode for Geographic Information Systems.

Later approaches suggested directly matching a sketch against other spatial representations, for example, a base map [45, 65]. In particular, this may be used for localization purposes, either of particular places or, indeed, the users themselves [46]. Landmarks are a common element in people's sketch maps [50,63]. With sketches drawn for localization, landmarks are even more likely to be crucial parts of the sketch because next to the configuration of the street network these are the geographic objects most likely to be remembered or perceived by the users. Given that landmarks are salient objects, they are also ideal candidates to restrict the search space. Thus, the ability to understand graphical landmark references is an important feature of spatial services employing a sketch-based interface.

6.4   Evaluating Landmark-Based Human-Computer Interaction

Throughout this book we have argued that incorporating landmarks is a necessity for truly intelligent spatial services. We have also argued that services producing and understanding landmarks deliver significant benefits for the human partner in human-computer interaction. In this section, we will provide some evidence for

---

[11]http://www.cyc.com/platform/opencyc, last visited 8/1/2014.

these arguments. We will discuss usability studies and other forms of evaluation that explicitly looked at the benefits of incorporating landmarks. A word of warning in the beginning though: not all is well yet in the realm of landmark-based services.

The following studies focus on aspects of landmark production. They are interested in how well (or how much better) users understand a service's communication output if this service employs landmarks. Most of them look at verbal production, but there is some work done on graphical landmark production as well.

Let us start with a requirement analysis. Burnett and his group have done a range of such studies as well as some work on human factors in landmark use (e.g., [8,51]) in the context of in-car and pedestrian navigation systems.

In the 1990s studies compared having a passenger providing instructions vs. using a car navigation system. A passenger who has detailed route knowledge and provides clear and timely instructions arguably presents the ideal situation. Compared to this, drivers using a navigation system made more navigation errors, took longer to complete a route, spent less time looking on the road or in the mirrors, rated their mental workload to be higher, and were rated by an expert to have lower quality of driving [9,27]. Based on these studies and other findings, Burnett [8] listed several reasons why navigation systems should include references to landmarks reflecting the arguments we have made: (1) Landmarks are consistent with basic human navigation strategies; (2) landmarks are valued by drivers; (3) landmarks are effective and efficient in navigation tasks; (4) landmarks increase user satisfaction.

Argument (1) has been widely discussed in the book. Regarding Argument (2), for example, a survey of 1,158 UK drivers found out that landmarks are the second most popular information type (after left-right information) that participants would want from their passengers helping in navigation [11]. Burnett's group found that participants either identified landmarks as crucial information in navigation situations or produced landmark references to support others in navigation tasks, depending on the condition tested and consistent with what we have discussed earlier [10, 51]. Arguments (3) and (4) will be discussed in more detail in the following—that is what this section is all about.

In an evaluation study testing pedestrian navigation instructions [59], participants either received instructions relying on distances and street names, or enhanced instructions with landmark references (similar to those of WhereIs). Ross et al. found that participants receiving enhanced instructions were significantly more confident in taking the right decisions at decision points and, indeed, also made fewer navigation errors compared to those receiving the basic instructions. Their study clearly makes a case for the inclusion of landmarks.

The Kiosk system [16] proved to be successful in guiding wayfinders as well. As already mentioned in Sect. 6.3.1 the rate of parsed utterances is rather low with only about 17 %. However, the keyword spotter manages to cope for this and detects useful keywords in 80 % of the utterances (leaving only 3 % of unparsed utterances).

In the Kiosk system evaluation, 26 participants, who were mostly unfamiliar with the environment, had to find their way to six different locations in a university building after negotiating with the kiosk how to get there. Given the high detection rate for keywords, only short dialogs were needed for users to receive the required information from the kiosk system [19]. This is an indicator for efficient dialog handling. More importantly, the overall user satisfaction was 90 %. The reason may be that almost 90 % of the participants found the test location eventually, and about 80 % of those with no or only small wayfinding problems (i.e., only minor confusion along the way, but without really taking wrong turns) [19].

Dethlefs et al. [20] evaluated their approach in a computer-based survey without actual wayfinding. Participants had to rate turn-by-turn route directions and destination descriptions generated by their approach, descriptions produced by a human direction giver, and by Google Maps (no destination descriptions from Google Maps; no commercial system is capable of producing them). Participants were asked to determine which of the presented instructions were automatically generated by a computer and which instructions appeared to be most useful. Instructions generated by Google Maps were clearly identified to be computer-based (94 % of participants). For the descriptions generated by Dethlefs et al.'s approach, only 36 % were classified as computer-generated, thus, 64 % of the participants took them as being from a human communication partner. This applies to turn-by-turn directions. Forty-two percent of the destination descriptions were correctly identified as computer-generated, but also 34 % of the human directions were falsely taken to be from a machine. In conclusion, this approach produces instructions that appear to be more natural than those generated by Google Maps, and may, thus, be better suited for human-computer interaction.

However, in terms of usefulness ratings results tell a different story. While destination descriptions by Dethlefs et al.'s approach are rated by both familiar and unfamiliar participants to be more useful than the human-generated ones (53 vs. 46 %; 65 vs. 33 %, respectively) turn-by-turn directions are only perceived as most useful by 7 % of the familiar participants. Human directions were seen to be most useful here, just ahead of those by Google Maps (48 vs. 42 %). Unfamiliar users are more positive; 37 % prefer Dethlefs et al.'s directions. Familiar users may reject them because the instructions tend to be verbose and, as discussed in Sect. 6.2.1, may include odd landmark references. The significant caveat with this study is that participants did not actually have to find their way. The study only tested for naturalness (where their approach is strong) and user preference (where it is popular with unfamiliar participants, but not so much with familiar ones). No actual performance data has been collected, therefore, no statements regarding which kind of directions is actually better can be made.

The SpaceBook system [38] was evaluated in a setup where participants had to perform eight tasks in two runs. These included both navigation and tourist information tasks. The system was tested against a baseline system relying on standard smartphone applications. Participants rated both systems equally successful in terms of task completion. However, the baseline system had a better task completion rate in most navigation tasks, whereas the SpaceBook system performed better in the tourist information tasks. This is also reflected in user preferences. The base system is preferred for navigation tasks, the SpaceBook system for tourist information tasks. With the SpaceBook system, users had major issues with navigation because the system did not provide any graphical information, i.e., no map or directional arrows.

Further, latency and positional errors of the GPS produced some directions too late or at the wrong locations, which harmed navigation ease and success.

The correct timing of verbal instructions for GPS-based navigation systems was addressed by Rehrl et al. [55]. They tested instructions using either metric information or landmark references to indicate where to turn. Participants navigated test routes in the city of Salzburg wearing headphones that canceled most of the street noise. An instructor shadowing the participants triggered instructions at pre-defined locations to avoid any timing issues that may result from poor GPS positioning. Similar to the SpaceBook project, these instructions were voice-only, no graphical route information was provided to the participants.

Overall, the study showed that the correct timing of unambiguous instructions is most crucial for successful wayfinding. Landmark references help removing ambiguity. While the type of instructions had no effect on walking time, participants made notably fewer navigation errors when using the landmark-based instructions compared to those receiving metric references. Basically, participants did not use metric information at all, but simply waited for the next instruction to come, whereas landmarks helped them to identify correct turns.

The system of Hile et al. [34] (see Sect. 6.2.2) combines photographic views along a route with verbal instructions. In their evaluation, participants could switch between a photo view and a map view. Most used both in their navigation, with the photo view mostly applied in critical, ambiguous situations. The participants perceived a range of the presented photographs as confusing, because in the real world trees blocked the view to the landmark depicted on the photo and referenced in the verbal instructions, and because photos did not coincide with the participants' perspective. In the latter case, participants had to mentally transform the perspective seen on the photograph to the perspective they had while moving, which is a cognitively demanding task. This clearly indicates that while graphical landmark references, and especially photographs, can be a powerful support for human users of a service, a careful selection of these photographs is crucial. Hile et al. addressed some of the issues identified by the study participants in some follow-up work [35].

Beeharee and Steed [4] got similar results from their user evaluation. Participants were clearly faster with photo-augmented instructions compared to only map views because they could use the photographs to disambiguate situations and to confirm that they were on the right track. But photos not taken directly from the participants' perspective, or in a different season, or different lighting conditions may confuse (some) users as again there is a perceptual mismatch that requires cognitive effort to resolve.

Wither et al. [68] compared their panorama view based navigation service (with and without enlarged business signs) to a 'traditional' map based navigation mode (again with or without enlarged business signs). They found no difference in navigation performance between the different modes. However, other than expected participants spent more time looking at the panorama view than on the map. Initially, Wither et al. hypothesized that panoramas were easier to match to the real world, since they essentially reproduce the perspective in the real world directly. But it seems that both increased visual complexity of panorama images and people's prior experience of using map based navigation counter this expectation. The enlarged signs had no effect on navigation performance, but were rated as a useful feature.

To sum up, essentially all studies presented in this section clearly demonstrate the advantages spatial information services gain from incorporating landmarks. But they also show that there are still a range of design challenges, many due to data issues (discussed in Sect. 5.5.1), issues of selecting truly relevant and identifiable landmarks (as exemplified in the study of Hile et al. [34]), i.e., of getting the context right (Chap. 4), and issues of finding a common ground in describing landmarks (as discussed in the beginning of Sect. 6.3.1).

6.5 Summary

In this chapter, we looked at landmarks in the interplay between human users and spatial information services. We highlighted some of the requirements for machine spatial communication. We then discussed in detail what it takes for a service to produce and understand landmark references, both verbally and graphically. In particular, we argued why it is easier to produce landmarks than to understand them. As in previous chapters, we also looked at some examples from research that aim at either of these tasks. Finally, we discussed a range of studies that demonstrate the power of landmark-based communication in human-computer interaction. This chapter now concludes our argument for why landmarks are a crucial element for truly intelligent spatial information systems. The results of the evaluation studies provide convincing evidence for our hypothesis. However, the studies also illustrate that there is still work to do.

References

1. Allen, G.L.: From knowledge to words to wayfinding: issues in the production and comprehension of route directions. In: Hirtle, S.C., Frank, A.U. (eds.) Spatial Information Theory. Lecture Notes in Computer Science, vol. 1329, pp. 363–372. Springer, Berlin (1997)
2. Bateman, J.A., Hois, J., Ross, R., Tenbrink, T.: A linguistic ontology of space for natural language processing. Artif. Intell. **174**(14), 1027–1071 (2010)
3. Beeharee, A., Steed, A.: Filtering location-based information using visibility. In: Strang, T., Linnhoff-Popien, C. (eds.) Location- and Context-Awareness. Lecture Notes in Computer Science, vol. 3479, pp. 306–315. Springer, Berlin (2005)
4. Beeharee, A.K., Steed, A.: A natural wayfinding exploiting photos in pedestrian navigation systems. In: Proceedings of the 8th Conference on Human-Computer Interaction with Mobile Devices and Services, MobileHCI '06, pp. 81–88. ACM, New York (2006)
5. Belz, A.: Automatic generation of weather forecast texts using comprehensive probabilistic generation-space models. Nat. Lang. Eng. **14**(4), 431–455 (2008)
6. Bittner, T., Stell, J.G.: Stratified rough sets and vagueness. In: Kuhn, W., Worboys, M.F., Timpf, S. (eds.) Spatial Information Theory. Lecture Notes in Computer Science, vol. 2825, pp. 270–286. Springer, Berlin (2003)

7. Blaser, A.D., Egenhofer, M.J.: A visual tool for querying geographic databases. In: AVI '00: Proceedings of the Working Conference on Advanced Visual Interfaces, pp. 211–216. ACM, New York (2000)

8. Burnett, G.: Turn right at the traffic lights: the requirement for landmarks in vehicle navigation systems. J. Navigation **53**(3), 499–510 (2000)

9. Burnett, G., Joyner, S.M.: An assessment of moving map and symbol-based route guidance systems. In: Noy, Y.I. (ed.) Ergonomics and Safety of Intelligent Driver Interfaces, pp. 115–136. Lawrence Erlbaum Associates, Mahwah (1997)

10. Burnett, G., Smith, D., May, A.: Supporting the navigation task: characteristics of 'good' landmarks. In: Hanson, M.A. (ed.) Contemporary Ergonomics 2001, pp. 441–446. Taylor and Francis, London (2001)

11. Burns, P.C.: Navigation and the older driver. Unpublished Ph.D. Thesis, Loughborough University (1997)

12. Burrough, P.A., Frank, A.U. (eds.): Geographic Objects with Indeterminate Boundaries. Taylor and Francis, London (1996)

13. Chipofya, M., Wang, J., Schwering, A.: Towards cognitively plausible spatial representations for sketch map alignment. In: Egenhofer, M.J., Giudice, N., Moratz, R., Worboys, M.F. (eds.) Spatial Information Theory. Lecture Notes in Computer Science, vol. 6899. Springer, Berlin (2011)

14. Clark, S., Hockenmaier, J., Steedman, M.: Building deep dependency structures with a wide-coverage CCG parser. In: Proceedings of the 40th Annual Meeting of the Association for Computational Linguistics, ACL '02, pp. 327–334. Association for Computational Linguistics, Stroudsburg (2002)

15. Couclelis, H.: Verbal directions for way-finding: space, cognition, and language. In: Portugali, J. (ed.) The Construction of Cognitive Maps. GeoJournal Library, vol. 32, pp. 133–153. Kluwer, Dordrecht (1996)

16. Cuayáhuitl, H., Dethlefs, N., Richter, K.F., Tenbrink, T., Bateman, J.: A dialogue system for indoor way-finding using text-based natural language. Int. J. Comput. Ling. Appl. **1**(1–2), 285–304 (2010)

17. Dale, R., Geldof, S., Prost, J.P.: Using natural language generation in automatic route description. J. Res. Pract. Inform. Tech. **37**(1), 89–105 (2005)

18. Denis, M.: The description of routes: a cognitive approach to the production of spatial discourse. Curr. Psychol. Cognit. **16**(4), 409–458 (1997)

19. Dethlefs, N., Cuayáhuitl, H., Richter, K.F., Andonova, E., Tenbrink, T., Bateman, J.: Evaluating task success in a dialogue system for indoor navigation. In: Lupkowski, P., Purve, M. (eds.) Aspects of Semantics and Pragmatics of Dialogue. SemDial 2010, pp. 143–146. Polish Society for Cognitive Science, Poznań (2010)

20. Dethlefs, N., Wu, Y., Kazerani, A., Winter, S.: Generation of adaptive route descriptions in urban environments. Spatial Cognition & Computation **11**(2), 153–177 (2011)

21. Douglas, D.H., Peucker, T.K.: Algorithms for the reduction of the number of points required to represent a digitized line or its caricature. Cartographica Int. J. Geogr. Inform. Geovisualization **10**(2), 112–122 (1973)

22. Duckham, M., Winter, S., Robinson, M.: Including landmarks in routing instructions. J. Location-Based Serv. **4**(1), 28–52 (2010)

23. Egenhofer, M.J.: Query processing in spatial-query-by-sketch. J. Vis. Lang. Comput. **8**, 403–424 (1997)

24. Egenhofer, M.J., Herring, J.R.: A mathematical framework for the definition of topological relationships. In: Brassel, K., Kishimoto, H. (eds.) 4th International Symposium on Spatial Data Handling, pp. 803–813. International Geographical Union, Zürich (1990)

25. Egenhofer, M.J., Mark, D.M.: Naive geography. In: Frank, A.U., Kuhn, W. (eds.) Spatial Information Theory. Lecture Notes in Computer Science, vol. 998, pp. 1–15. Springer, Berlin (1995)

26. Elias, B., Paelke, V., Kuhnt, S.: Concepts for the cartographic visualization of landmarks. In: Gartner, G. (ed.) Location Based Services and Telecartography: Proceedings of the Symposium 2005, Geowissenschaftliche Mitteilungen, pp. 1149–1155. TU Vienna, Vienna (2005)

27. Fastenmeier, W., Haller, R., Lerner, G.: A preliminary safety evaluation of route guidance comparing different MMI concepts. In: Proceedings of the First World Congress on Applications of Transport Telemetrics and Intelligent Vehicle Highway Systems, vol. 4, pp. 1750–1756. Artech House, Boston (1994)

28. Forbus, K.D., Ferguson, R.W., Usher, J.M.: Towards a computational model of sketching. In: Proceedings of the 6th International Conference on Intelligent User Interfaces, IUI '01, pp. 77–83. ACM, New York (2001)

29. Forbus, K.D., Usher, J., Lovett, A., Lockwood, K., Wetzel, J.: Cogsketch: sketch understanding for cognitive science research and for education. Top. Cognit. Sci. 3(4), 648–666 (2011)

30. Frank, A.U.: Qualitative spatial reasoning about distances and directions in geographic space. J. Vis. Lang. Comput. 3, 343–371 (1992)

31. French, R.M.: Moving beyond the turing test. Comm. ACM 55(12), 74–77 (2012)

32. Goodman, J., Gray, P., Khammampad, K., Brewster, S.: Using landmarks to support older people in navigation. In: Brewster, S., Dunlop, M. (eds.) Mobile Human-Computer Interaction: MobileHCI 2004. Lecture Notes in Computer Science, vol. 3160, pp. 38–48. Springer, Berlin (2004)

33. Grice, P.: Logic and conversation. Syntax and Semantics 3, 41–58 (1975)

34. Hile, H., Vedantham, R., Cuellar, G., Liu, A., Gelfand, N., Grzeszczuk, R., Borriello, G.: Landmark-based pedestrian navigation from collections of geotagged photos. In: Proceedings of the 7th International Conference on Mobile and Ubiquitous Multimedia, MUM '08, pp. 145–152. ACM, New York (2008)

35. Hile, H., Grzeszczuk, R., Liu, A., Vedantham, R., Košecka, J., Borriello, G.: Landmark-based pedestrian navigation with enhanced spatial reasoning. In: Tokuda, H., Beigl, M., Friday, A., Brush, A., Tobe, Y. (eds.) Pervasive Computing. Lecture Notes in Computer Science, vol. 5538, pp. 59–76. Springer, Berlin (2009)

36. Hill, L.L.: Georeferencing: The Geographic Associations of Information. Digital Libraries and Electronic Publishing. MIT Press, Cambridge (2006)

37. Hirtle, S.C., Sorrows, M.E.: Designing a multi-modal tool for locating buildings on a college campus. J. Environ. Psychol. 18(3), 265–276 (1998)

38. Janarthanam, S., Lemon, O., Bartie, P., Dalmas, T., Dickinson, A., Liu, X., Mackaness, W., Webber, B.: Evaluating a city exploration dialogue system combining question-answering and pedestrian navigation. In: Proceedings of the 51st Annual Meeting of the Association of Computational Linguistics, pp. 1660–1668. Sofia, Bulgaria (2013)

39. Jurafsky, D., Martin, J.H.: Speech and Language Processing: An Introduction to Natural Language Processing, Computational Linguistics, and Speech Recognition, 2nd edn. Pearson Prentice Hall, Upper Saddle River (2008)

40. Klippel, A., Montello, D.R.: Linguistic and non-linguistic turn direction concepts. In: Winter, S., Duckham, M., Kulik, L., Kuipers, B. (eds.) Spatial Information Theory. Lecture Notes in Computer Science, vol. 4736, pp. 354–372. Springer, Berlin (2007)

41. Klippel, A., Tappe, H., Habel, C.: Pictorial representations of routes: chunking route segments during comprehension. In: Freksa, C., Brauer, W., Habel, C., Wender, K.F. (eds.) Spatial Cognition III. Lecture Notes in Artificial Intelligence, vol. 2685, pp. 11–33. Springer, Berlin (2003)

42. Klippel, A., Richter, K.F., Hansen, S.: Structural salience as a landmark. In: Workshop Mobile Maps 2005. Salzburg, Austria (2005)

43. Klippel, A., Hansen, S., Richter, K.F., Winter, S.: Urban granularities: a data structure for cognitively ergonomic route directions. GeoInformatica 13(2), 223–247 (2009)

44. Kolbe, T.H.: Augmented videos and panoramas for pedestrian navigation. In: Gartner, G. (ed.) Proceedings of the 2nd Symposium on Location Based Services and TeleCartography, Geowissenschaftliche Mitteilungen. TU Vienna, Vienna (2004)

45. Kopczynski, M.: Efficient spatial queries with sketches. In: Proceedings of the ISPRS Technical Commission II Symposium, pp. 19–24, Vienna, Austria (2006)
46. Kopczynski, M., Sester, M.: Graph based methods for localisation by a sketch. In: Proceedings of the 22nd International Cartographic Conference (ICC2005). La Coruna, Spain (2005)
47. Kulik, L.: A geometric theory of vague boundaries based on supervaluation. In: Montello, D.R. (ed.) Spatial Information Theory. Lecture Notes in Computer Science, vol. 2205, pp. 44–59. Springer, Berlin (2001)
48. Lee, Y., Kwong, A., Pun, L., Mack, A.: Multi-media map for visual navigation. J. Geospatial Eng. 3(2), 87–96 (2001)
49. Lovelace, K.L., Hegarty, M., Montello, D.R.: Elements of good route directions in familiar and unfamiliar environments. In: Freksa, C., Mark, D.M. (eds.) Spatial Information Theory. Lecture Notes in Computer Science, vol. 1661, pp. 65–82. Springer, Berlin (1999)
50. Lynch, K.: The Image of the City. The MIT Press, Cambridge (1960)
51. May, A.J., Ross, T., Bayer, S.H., Tarkiainen, M.J.: Pedestrian navigation aids: Information requirements and design implications. Pers. Ubiquit. Comput. 7(6), 331–338 (2003)
52. Montello, D.R., Frank, A.U.: Modeling directional knowledge and reasoning in environmental space: testing qualitative metrics. In: Portugali, J. (ed.) The Construction of Cognitive Maps. GeoJournal Library, vol. 32, pp. 321–344. Kluwer, Dordrecht (1996)
53. Porzel, R., Gurevych, I., Malaka, R.: In context: Integrating domain- and situation-specific knowledge. In: Wahlster, W. (ed.) SmartKom: Foundations of Multimodal Dialogue Systems, pp. 269–284. Springer, Berlin (2006)
54. Raubal, M., Winter, S.: Enriching wayfinding instructions with local landmarks. In: Egenhofer, M.J., Mark, D.M. (eds.) Geographic Information Science. Lecture Notes in Computer Science, vol. 2478, pp. 243–259. Springer, Berlin (2002)
55. Rehrl, K., Häusler, E., Leitinger, S.: Comparing the effectiveness of GPS-enhanced voice guidance for pedestrians with metric- and landmark-based instruction sets. In: Fabrikant, S., Reichenbacher, T., van Kreveld, M., Schlieder, C. (eds.) Geographic Information Science. Lecture Notes in Computer Science, vol. 6292, pp. 189–203. Springer, Berlin (2010)
56. Richter, K.F.: A uniform handling of different landmark types in route directions. In: Winter, S., Duckham, M., Kulik, L., Kuipers, B. (eds.) Spatial Information Theory. Lecture Notes in Computer Science, vol. 4736, pp. 373–389. Springer, Berlin (2007)
57. Richter, K.F.: Context-Specific Route Directions: Generation of Cognitively Motivated Wayfinding Instructions, vol. DisKi 314 / SFB/TR 8 Monographs Volume 3. IOS Press, Amsterdam (2008)
58. Richter, K.F., Winter, S., Santosa, S.: Hierarchical representations of indoor spaces. Environ. Plann. B Plann. Des. 38(6), 1052–1070 (2011)
59. Ross, T., May, A., Thompson, S.: The use of landmarks in pedestrian navigation instructions and the effects of context. In: Brewster, S., Dunlop, M. (eds.) Mobile Computer Interaction: MobileHCI 2004. Lecture Notes in Computer Science, vol. 3160, pp. 300–304. Springer, Berlin (2004)
60. Takacs, G., Chandrasekhar, V., Gelfand, N., Xiong, Y., Chen, W.C., Bismpigiannis, T., Grzeszczuk, R., Pulli, K., Girod, B.: Outdoors augmented reality on mobile phone using loxel-based visual feature organization. In: Proceedings of the 1st ACM International Conference on Multimedia Information Retrieval, MIR '08, pp. 427–434. ACM, New York (2008)
61. Tomko, M., Winter, S., Claramunt, C.: Experiential hierarchies of streets. Comput. Environ. Urban Syst. 32(1), 41–52 (2008)
62. Turing, A.M.: Computing machinery and intelligence. Mind 59(236), 433–460 (1950)
63. Tversky, B., Lee, P.U.: Pictorial and verbal tools for conveying routes. In: Freksa, C., Mark, D.M. (eds.) Spatial Information Theory. Lecture Notes in Computer Science, vol. 1661, pp. 51–64. Springer, Berlin (1999)
64. Vasardani, M., Winter, S., Richter, K.F.: Locating place names from place descriptions. Int. J. Geogr. Inform. Sci. 27(12), 2509–2532 (2013)

65. Wallgrün, J.O., Wolter, D., Richter, K.F.: Qualitative matching of spatial information. In: Proceedings of the 18th SIGSPATIAL International Conference on Advances in Geographic Information Systems, GIS '10, pp. 300–309. ACM, New York (2010)

66. Winter, S.: Spatial intelligence: ready for a challenge? Spatial Cognit. Comput. 9(2), 138–151 (2009)

67. Winter, S., Truelove, M.: Talking about place where it matters. In: Raubal, M., Mark, D.M., Frank,A.U. (eds.) Cognitive and Linguistic Aspects of Geographic Space: New Perspectives on Geographic Information Research. Lecture Notes in Geoinformation and Cartography, pp. 121–139. Springer, Berlin (2013)

68. Wither, J., Au, C.E., Rischpater, R., Grzeszczuk, R.: Moving beyond the map: Automated landmark based pedestrian guidance using street level panoramas. In: Proceedings of the 15th International Conference on Human-Computer Interaction with Mobile Devices and Services, MobileHCI, pp. 203–212. ACM, New York (2013)

69. Worboys, M., Duckham, M., Kulik, L.: Commonsense notions of proximity and direction in environmental space. Spatial Cognit. Comput. 4(4), 285–312 (2004)

70. Zhang, Q.: Multi-modal landmark-integrated route instructions. Unpublished Masters Thesis, Department of Infrastructure Engineering, The University of Melbourne (2012)

Chapter 7
Conclusions: What Is Known and What Is Still Challenging About Landmarks

Abstract This chapter concludes the book. It briefly summarizes what we have discussed in the previous six chapters and then looks ahead. In particular, we contemplate what it takes for a geospatial system to be *intelligent*, and what we still miss at the moment in order to build such systems. Overall, we believe that we have provided an appreciation and better understanding of both the challenges and potential of landmarks in intelligent geospatial systems.

7.1 What We Know: A Short Summary

This book set out to summarize the current state of knowledge about landmarks largely based on cognitive research, and how this makes geo-spatial systems more capable of interacting with human beings.

We started with defining what a landmark actually is, and very soon realized that this is not a trivial task. We believe that it is crucial to relate the concept of landmarks directly to people's embodied experience and their cognitive processing of their living environments and, thus, define landmarks to be *geographic objects that structure human mental representations of space*. They serve as anchor points in our mental representations; their internal structure is not important in that respect. This sets landmarks apart from other geographic concepts that seem similar, such as places or points of interest.

We have seen that there is a large body of research in the cognitive and neurosciences backing this definition. Landmarks serve as anchor and reference points in our mental representations and communication, which makes them crucial in our understanding of our world. They do this because landmarks stick out. Each landmark has some aspects that makes it grab our attention. These aspects depend on the context in which we encounter a potential landmark.

If we want computers to use landmarks in their interaction with us, we need to somehow make that elusive concept landmark known to them. We have exemplified this by presenting an—admittedly—simple algebraic formalization, but

K.-F. Richter and S. Winter, *Landmarks: GIScience for Intelligent Services*,
DOI 10.1007/978-3-319-05732-3_7, © Springer International Publishing Switzerland 2014

more importantly we have established that being a landmark is a property of a geographic object. There are no objects of type landmark. Consequently, computational approaches to using landmarks need to incorporate means to determine whether specific geographic objects may grab our attention. First, a computational system has to calculate which geographic objects are *salient*, i.e., may in principle serve as a landmark candidate. From all these candidates, the system then has to select the most *relevant* one for any given situation. We termed these two processes *landmark identification* and *landmark integration*. Various approaches for both these processes have been developed in the past 10 years or so, however, most fall short when it comes to integrating both processes. More importantly, these approaches usually have unrealistic demands on the data sets they rely on—today's infrastructure is simply not able to provide sufficiently detailed and evenly distributed data on landmarks.

Finally, we have looked at the interaction between human and computer with specific focus on orientation and wayfinding. All approaches we have discussed clearly show the importance and benefits of landmark references in this interaction. However, they also highlight some fundamental issues that still wait for satisfying solutions.

Thus, while landmarks clearly promise major benefits in human-computer interaction, and although tremendous progress has been made over the years in incorporating them into geospatial systems, there are still challenges ahead before truly intelligent systems emerge.

7.2   What We Aim for: Intelligent Geospatial Systems

Remember HAL, the artificial intelligence that controls the spacecraft and interacts with the astronauts in Stanley Kubrick's film *2001: A Space Odyssey*? HAL was capable of remarkable communication skills, including learning mechanisms, some of them so creepy as to create the suspense in the movie. For example, note this competence of introspection: "I know I've made some very poor decisions recently, but I can give you my complete assurance that my work will be back to normal. I've still got the greatest enthusiasm and confidence in the mission. And I want to help you." That was way back in 1968. And it was fiction.

How would HAL have been communicating with the astronauts on spatial tasks, such as orientation and wayfinding? To limit context and expectations, let us stick with terrestrial environments (and abandon the creepiness of HAL). How would an *intelligent* system guide us through traffic, to our hotel, remind us where we would have a meeting tomorrow or where we have parked our car last night?

We believe such systems should support human interaction with the environment, but not control this interaction. After all, Alan Turing set as the gauge for an intelligent system a machine that is capable of communicating like a person, not one that functions like a person [17]. Not more and not less. But we have also pointed out several times in this book that the goal for intelligent geospatial systems needs to be to also make up for human mistakes and cognitive limitations. Systems interacting with humans in a way meaningful to the human seems to be a reasonable expectation for an intelligent geospatial system. It is also an expectation more inspiring than the pure imitation of people's sometimes unreliable spatial skills and spatial communication behavior.

At this stage we should clarify our own aspirations in using computers to understand human cognition and in using them to achieve artificial cognition. In line with French's call for meaningful interaction [7], this book does search to enable the computer to understand human spatial expressions, and to generate spatial expressions that can easily be digested by people. The suggested and reviewed formal models do aim to enter into a meaningful dialog with a person, and hence, need to sufficiently understand human spatial cognition. But they do not aspire to explain and map processes of human spatial cognition and communication behavior on representations in a computer. This book is not about cognitive modeling.

In other words we do not see a need for computational processes replicating the neurological basis of spatial representations in the brain in order to achieve a meaningful dialog supporting a person in spatial decision making. But we do see a need for models of landmarks that capture the nature of landmarks and are able to relate to human embodied experiences.

7.3   What We Need to Know: Human-Computer Interaction

We see two major obstacles why modeling landmarks in computational systems is still a great challenge. We believe that these two would definitely need to be overcome in order to develop the kind of intelligent geospatial systems we have just described.

1. Knowledge: The computer does not have embodied experiences of physical reality.

   In this book, much of the human embodied experience as it has been identified by cognitive science research has been taken as basis for formal models, differing from current approaches based on directories or gazetteers. Lacking embodied experiences, a computational system has to find some substitute landmark 'experience', i.e., smart ways of reasoning about existing data about an environment. One such approach will always be machine learning from large datasets, a stochastical approach successfully demonstrated among others by IBM's Watson computer. Using the same philosophy, the Todai Robot[1] sets out to pass the entry exams of Japan's leading university, which includes questions requiring spatial abilities such as predicting trajectories. Factual spatial knowledge, however, would be sought from the data resources of the web or large databases. Another

---

[1]http://21robot.org/about/?lang=english, last visited 3/1/2014.

feasible approach uses (spatial) reasoning mechanisms to identify landmarks based on, among others, a geographic object's location. Likely, in the end a combination of these two approaches will prove most successful.

Independent of the chosen approach, having the right data available might ultimately be the greatest challenge. In Chaps. 5 and 6, we have seen that today's approaches rely on detailed data being available for all potential landmark candidates across an environment. This is an unrealistic demand, even for an approach such as the one by Duckham et al. [5] that only needs type information for POIs.

2. Context: No system has yet a grasp of context in any way comparable to a person.

Face-to-face communication includes a multitude of often subconscious and automated considerations regarding the partner's physiognomy, mood, or other stimulations of visual, auditory, tactile or other senses. People are highly skilled in non-verbal communication, and constantly make inferences about context without recognizable effort. Machines have a very limited sensory input, which is one reason why they have a hard time with non-verbal communication—think about reading your communication partner's mood in a video conference call with poor connection.

But even already the verbal (or graphical) communication is challenging for a computer. Understanding the semantics of language and making correct inferences is one of the big research challenges in artificial intelligence. Language is under-specified and hence ambiguous. We have seen examples of this challenge, such as the ambiguity of the reference frame for the interpretation of spatial relationships. Does "Melbourne Cup" mean a race carried out in Melbourne, or is it a proper name? Does "left of the school" mean *left* in the allocentric system of the school, or in the egocentric system of the person providing this utterance while standing in front of the school? If somebody is looking for a kindergarten near their home, is that *near* the same as in "the café near the library"? As a final example, and referring to the mentioned contrast sets [18], is the spatial granularity of the references to the Eiffel Tower the same in "Let's meet on the observation deck of the Eiffel Tower" and "I have visited Notre Dame, Louvre and the Eiffel Tower"? A person would interpret these ambiguous messages by considering the given communication context and some default reasoning [10,19], while the computer would stick with term recognition, and perhaps some stochastic interpretations from machine learning.

In terms of providing knowledge for our intelligent geospatial system we envision a combination of several different approaches using traditional data sets, data mining of 'neo-geography' sources, and user-generated content platforms.

Google Street View or Nokia City Scene have increasing coverage—such panorama views of the environment may be the closest a system may ever get to an embodied experience. While automatically identifying signage and shop labels in these images is an impressive achievement, it is only the first step, and easier than, say, automatically identifying churches or bus stops. Even more challenging is identifying the 'outstanding' building in a given scene. Remember the grey world assumption we have discussed in Chap. 5 (on p. 139). We have seen that finding outstanding geographic objects may be supported by data mining methods (Sect. 5.2.2), and reasoning about object types (Sects. 5.3 and 5.5.2).

A combination of these approaches will likely get us quite far in identifying landmarks in densely populated areas, in particular in city centers and commercial districts. However, given the uneven distribution of POI data and the lack of business signage in residential suburbs, there are also large areas of our environment where we would like an intelligent system to refer to landmarks, but it fails to do so for lack of data. This is where user-generated content steps in. Implementing a platform that allows users to contribute landmark information about their neighborhood would be an ideal supplement to the machine learning techniques used for an automatic identification of landmarks. Such a platform must be easy to use, allow for quick rewards for the contributor, but it also needs to ensure that the contributed data can be applied in landmark identification approaches in a straightforward manner [12]. We see the OpenLandmarks platform as a first step in the right direction (see Sect. 5.5.3). A database such a user-generated content platform feeds into may be pre-filled by all landmark candidates we can get from the data mining methods discussed before. This would serve two purposes: first, contributors are not faced with a blank canvas, but can already see what their contributions will (are supposed to) look like. Second, the end result will be an integrated database that combines user-generated content and automatically extracted landmarks. This will also allow for correction mechanisms. Users may change information on landmarks the system identified, and likewise the system may be able to flag improbable landmark candidates submitted by a user.

It may well be the case that the success of such a user-generated content approach to collecting landmark information depends on the developed platform having a monopoly in collecting such data. Wikipedia would not be so successful if there were six or seven other online encyclopedias of similar status competing for contributors. The same holds for OpenStreetMap. OSM data would be much less useful if every country (or even only every continent) had their own platform for collecting topographic data, each with slightly different mechanisms and data structures. In fact, a recent change of the OSM license, which also affects the kind of data you can use as basis for your contributions, has led several OSM contributors to fork OSM data into a new data set where the old license still holds—this move has been particularly strong in Australia. The consequences of this fork are still to be seen, but some fragmentation and inconsistencies between the two data sets are most likely to occur.

Context is an altogether different beast. Context and its operationalization has been researched for a long time in human-computer interaction and location based

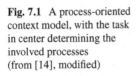

Fig. 7.1 A process-oriented context model, with the task in center determining the involved processes (from [14], modified)

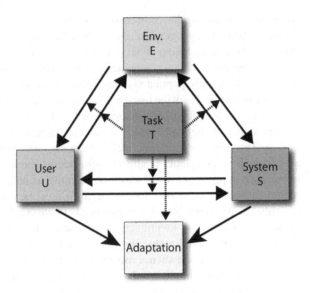

services (for a review of early work see [2]). The often cited definition by Dey [4] takes context to be "any information that can be used to characterize the situation of entities that are considered relevant to the interaction between a user and an application, including the user and the application themselves." This definition is certainly correct, but it is also similarly difficult to operationalize as some of the definitions of landmark that we have looked at in the beginning of this book. Thus, rather than compiling and parametrizing a list of factors that may describe the current context, which will never be complete, we propose a process-oriented view [6, 14].

This view on context is centered on the task at hand and focuses on the processes that occur between the involved entities. In the case of intelligent geospatial systems these entities are the system (S), the user (U), and the environment (E) that user and system act in and communicate about. Figure 7.1 provides a conceptual overview of such a context model. The model assumes goal-directed behavior by both user and system, and accordingly the task determines the processes performed by both of them. It also accounts for the influences different environments may have on task performance. Crucially, both user and system may need to adapt to the current context, i.e., processes of adaptation are a fundamental property of this model.

Accordingly, we suggest the following road ahead. The basic first step is to continue and refine the automatic production of skeletal route and place descriptions [3]. Or, more precisely—because skeletal descriptions are not formally defined—the aim is to produce descriptions of sufficient pragmatic information content that use a minimal number of references, i.e., descriptions that are relevant and short.

We have already come a long way here if you think of the substantial preliminary work both in verbal and graphical communication we discussed in this book (e.g., [1,8,11,16]). This work largely assumes an unfamiliar communication partner (the U in Fig. 7.1), who is the easiest to model. Other factors of the communication context have not yet been fully addressed, for example, an adaptation to a particular environment or the time of the day (E).

Another change to the environment, and consequently to the task (T), would be an adaptation of skeletal descriptions to emergency situations. Again, this is a change to context that is relatively easy to capture and model. It demands particularly clear and unambiguous communication. When people are under stress, there is no time for thinking about what the system may have actually meant.

Adapting descriptions to the prior knowledge of the user would be a desirable achievement. Such adaptation is fundamental for pragmatic information content, i.e., communicating only what is relevant. It would prevent your car navigation system from telling you how to get from your home to the highway in all detail—something you have likely done hundreds of times before. There is preliminary work done in that direction as well, most of which assumes that you own the system that communicates with you. These systems infer your knowledge by tracking your movements (e.g., [9, 15]). More challenging, but also more interesting, is a scenario where the system has to figure out the user's prior knowledge by dialog [13]. Such a scenario requires flexibility and adaptation, and also strategies of negotiation. This step gets us finally to (landmark) understanding, which we have argued to be more difficult than (landmark) producing.

These dialog-based systems will need to understand place descriptions by people in-situ, using location information as well as keeping track of what has been mentioned before in the current dialog. But it will require more than this. The system will have to be capable of mapping the user's understanding of an environment to its own representation. This means that the systems' internal representations will need to be created based on the defining elements of human spatial representation—landmarks.

In the long run these developments will lead to intelligent geospatial systems—systems as flexible and adaptive as a human communication partner, with re- and pro-active behavior and constructive communication. Landmarks are the elements that can tie it all together by structuring both the human and system's representation of space.

References

1. Agrawala, M., Stolte, C.: Rendering effective route maps: Improving usability through generalization. In: SIGGRAPH 2001, pp. 241–250. ACM, Los Angeles (2001)
2. Chen, G., Kotz, D.: A survey of context-aware mobile computing research. Technical Report TR2000-381. Dartmouth College, Hanover (2000)
3. Denis, M.: The description of routes: a cognitive approach to the production of spatial discourse. Curr. Psychol. Cognit. 16(4), 409–458 (1997)
4. Dey, A.K.: Understanding and using context. Pers. Ubiquit. Comput. 5(1), 4–7 (2001)
5. Duckham, M., Winter, S., Robinson, M.: Including landmarks in routing instructions. J. Location-Based Serv. 4(1), 28–52 (2010)

6. Freksa, C., Klippel, A., Winter, S.: A cognitive perspective on spatial context. In: Cohn, A., Freksa, C., Nebel, B. (eds.) Spatial Cognition: Specialization and Integration. Dagstuhl Seminar Proceedings, vol. 05491. Dagstuhl, Germany (2007)
7. French, R.M.: Moving beyond the Turing test. Comm. ACM **55**(12), 74–77 (2012)
8. Kopf, J., Agrawala, M., Bargeron, D., Cohen, M.F.: Automatic generation of destination maps. In: SIGGRAPH Asia, pp. 158:1–158:12. ACM, New York (2010)
9. Patel, K., Chen, M.Y., Smith, I., Landay, J.A.: Personalizing routes. In: UIST '06: Proceedings of the 19th Annual ACM Symposium on User Interface Software and Technology, pp. 187–190. ACM Press, New York (2006)
10. Rauh, R., Hagen, C., Knauff, M., Kuss, T., Schlieder, C., Strube, G.: Preferred and alternative mental models in spatial reasoning. Spatial Cognit. Comput. **5**(2–3), 239–269 (2005)
11. Richter, K.F.: Context-Specific Route Directions: Generation of Cognitively Motivated Wayfinding Instructions, vol. DisKi 314 / SFB/TR 8 Monographs Volume 3. IOS Press, Amsterdam (2008)
12. Richter, K.F., Winter, S.: Citizens as database: conscious ubiquity in data collection. In: Pfoser, D., Tao, Y., Mouratidis, K., Nascimento, M., Mokbel, M., Shekhar, S., Huang, Y. (eds.) Advances in Spatial and Temporal Databases. Lecture Notes in Computer Science, vol. 6849, pp. 445–448. Springer, Berlin (2011)
13. Richter, K.F., Tomko, M., Winter, S.: A dialog-driven process of generating route directions. Comput. Environ. Urban Syst. **32**(3), 233–245 (2008)
14. Richter, K.F., Dara-Abrams, D., Raubal, M.: Navigating and learning with location based services: a user-centric design. In: Gartner, G., Li, Y. (eds.) Proceedings of the 7th International Symposium on LBS and Telecartography, pp. 261–276 (2010)
15. Schmid, F.: Knowledge based wayfinding maps for small display cartography. J. Location Based Syst. **2**(1), 57–83 (2008)
16. Tomko, M., Winter, S.: Pragmatic construction of destination descriptions for urban environments. Spatial Cognit. Comput. **9**(1), 1–29 (2009)
17. Turing, A.M.: Computing machinery and intelligence. Mind **59**(236), 433–460 (1950)
18. Winter, S., Freksa, C.: Approaching the notion of place by contrast. J. Spatial Inform. Sci. **2012**(5), 31–50 (2012)
19. Winter, S., Truelove, M.: Talking about place where it matters. In: Raubal, M., Mark, D.M., Frank, A.U. (eds.) Cognitive and Linguistic Aspects of Geographic Space: New Perspectives on Geographic Information Research. Lecture Notes in Geoinformation and Cartography, pp. 121–139. Springer, Berlin (2013)

Author Index

A

Able, Kenneth P., 71
Acredolo, Linda P., 44, 67
Adam, Hartwig, 148
Agrawala, Maneesh, 210
Alberts, Denise M., 56
Allen, Gary L., 44, 48, 51, 56, 73, 82, 91, 178
Anderson, Anne H., 78
Andonova, Elena, 197
Anwar, Mohd, 168
Appleyard, Donald, 11, 56, 73
Aristotle, 14, 15
Ashburner, John, 63
Au, Carmen E., 187, 198
Aurenhammer, Franz, 124, 145
Austin, John Langshaw, 81

B

Backstrom, Lars, 149
Bader, Miles, 78
Bafna, Sonit, 122
Baldwin, Tim, 168
Bard, Ellen Gurman, 78
Bargeron, David, 210
Barkowsky, Thomas, 95
Bartie, Phil, 184, 191, 197
Bateman, John, 95, 181, 188, 190, 196, 197
Battestini, Agathe, 83
Battista, Christian, 30
Baumann, Oliver, 64
Bayer, Steven H., 196
Beall, Andrew C., 29, 157
Becker, Suzanna, 10
Beeharee, Ashweeni Kumar, 187, 198
Bell, Marek, 168
Bellgrove, Mark A., 64

Belz, Anja, 183
Benedikt, Michael L., 122
Benjamins, V. Richard, 112
Benner, Jessica G., 168
Berman, Dafna, 49
Bertolo, Laura, 59
Bertolo, Laura, 87, 89, 90
Bethus, Ingrid, 63
Bialystok, Ellen, 44
Billen, Roland, 111
Bismpigiannis, Thanos, 187
Bissaco, Alessandro, 148
Bittner, Steffen, 114
Bittner, Thomas, 114, 189
Blades, Mark, 48, 68, 84, 87
Blaser, Andreas D., 193, 194
Bodenheimer, Bobby, 47
Borriello, Gaetano, 187, 198, 199
Both, Alan, 27
Boye, Johan, 167
Boyes-Braem, Penny, 5, 6, 82
Boyle, Elizabeth, 78
Braddick, Oliver J., 65, 86
Braitenberg, Valentino, 27
Branaghan, Russell J., 49
Braun, Allen R., 94
Brennan, Penny L., 74
Brewster, Stephen, 187
Briggs, Ronald, 53, 125
Brook-Rose, Christine, 81
Brosset, David, 59
Brown, Barry, 168
Brown, John Seely, 21
Brucher, Fernando, 148
Brun, Vegard H., 62
Buchsbaum, Gershon, 139
Buddemeier, Ulrich, 148

K.-F. Richter and S. Winter, *Landmarks: GIScience for Intelligent Services*,
DOI 10.1007/978-3-319-05732-3, © Springer International Publishing Switzerland 2014

Subject Index

A

advance visibility, 141–143, 151, 157
affordance, 55
after-region, 154, 155
anchor point, 9, 10, 15, 23
artificial intelligence, 20, 190, 208

B

base level theory, 82
before-region, 154, 155
Bing Maps, 147
Bremen kiosk system, 181–185, 190, 191, 196

C

categories, 5, 156, 165, 166, 183
category resemblance, 5, 82
clustering, 144, 146, 149, 150, 187
    hierarchical agglomerative, 147, 148
Cobweb, 144
cognitive economy, 5
cognitive efficiency, 34, 71–73, 76, 124
cognitive map, 9, 61, 71
cognitive science, 42
cognitively ergonomic, 130, 131
CogSketch system, 195
configuration, 10
context, 6–8, 12, 13, 15–18, 21, 23, 29, 32, 36,
    48, 50, 59, 60, 62, 63, 74, 77, 81–85, 88,
    89, 91, 93, 95–97, 109–113, 115–123,
    125–132, 140, 144, 147, 165, 168, 176,
    178, 189, 190, 199, 205, 208–211
context awareness, 10, 111, 113
contrast, 15
CORAL system, 155, 180–184
crowdsourcing, 167, 187

D

description
    destination, 183, 184, 197
    place, 18, 29, 83, 87, 92–94, 126, 127, 189,
        210, 211
    route, 12, 18, 21, 48, 51, 55, 58, 59, 65,
        87–92, 95, 116, 128, 130, 131, 146, 155,
        158, 167, 177, 179, 181, 183, 184, 196,
        197, 210
destination, 31

E

embodied experience, 1, 7, 10, 13, 19, 21, 29,
    31, 36, 43, 53, 54, 66, 73, 77, 80, 94
etymology, 3
Exif, 149
experience, 207
EyeSpy, 168

F

fMRI, 61
function
    in wayfinding, 165

G

gender differences, 44
geographic information retrieval, 146
Gestalt, 78
global landmark, 31, 35–37, 56, 57, 71, 73, 74,
    116
Google Maps, 189, 197
Google Street View, 187, 208
graph, 35, 150, 152, 159, 160, 162, 166
    complete line, 158, 159
grey world assumption, 54, 139, 209

K.-F. Richter and S. Winter, *Landmarks: GIScience for Intelligent Services*,
DOI 10.1007/978-3-319-05732-3, © Springer International Publishing Switzerland 2014

Printed in the United States
By Bookmasters